Karl Popper's
Philosophy of Science

Routledge Studies in the Philosophy of Science

Karl Popper's Philosophy of Science

Rationality Without Foundations

Stefano Gattei

Routledge
Taylor & Francis Group
New York   London

First published 2009
by Routledge
270 Madison Ave, New York, NY 10016

Simultaneously published in the UK
by Routledge
2 Park Square, Milton Park, Abingdon, Oxon OX14 4RN

First issued in paperback 2010

Routledge is an imprint of the Taylor & Francis Group, an informa business

© 2009 Stefano Gattei

Typeset in Sabon by IBT Global.

Library of Congress Cataloging in Publication Data
Gattei, Stefano.
  Karl Popper's philosophy of science : rationality without foundations / by Stefano Gattei.
    p. cm. — (Routledge studies in the philosophy of science ; 5)
  Includes bibliographical references and index.
  ISBN-13: 978-0-415-37831-4 (hbk : alk. paper)
  ISBN-10: 0-415-37831-1 (hbk : alk. paper)
  1. Popper, Karl R. (Karl Raimund), 1902–1994.   2. Knowledge, Theory of.
3. Science—Philosophy.   I. Title.
  B1649.P64G38 2009
  121.092—dc22
  2008021890

'El Principio' from ATLAS Copyright © 1995, Maria Kodama. Reproduced by permission. English translation by Stefano Gattei. All rights reserved.

ISBN 978–0–415–37831-4 (hbk)
ISBN 978–0–415–88776-2 (pbk)
ISBN 978–0–203–88719-6 (ebk)

For my mother and father,
with gratitude and love

Two Greeks converse: Socrates and Parmenides, perhaps.
We may never know their names; the story, thus, will be more
                                        [mysterious and tranquil.
The theme of the dialogue is abstract. At times they allude to
                                        [myths, which they both distrust.
The reasons they advance may abound in fallacies and have no end.
They do not quarrel. And they wish neither to persuade nor to be
                                        [persuaded; they think of neither win nor loss.
They agree on one single thing: they know that discussion is the not-
                                        [impossible path to reach a truth.
Free of myth and metaphor, they think or try to.
We shall never know their names.
This conversation of two unknowns somewhere in Greece is the
                                        [capital event in History.
They forgot prayer and magic.

—Jorge Luis Borges "The Beginning", 1984

Contents

Abbreviations

ALPS	*All Life is Problem-Solving*, 1994/1999
BGE	*Die beiden Grundprobleme der Erkenntnistheorie*, 1979, 1994[2]
CR	*Conjectures and Refutations*, 1963, 1989[5]
FS	*Frühe Schriften*, 2006
KBMP	*Knowledge and the Body-Mind Problem*, 1994
LF	*Logik der Forschung*, 1935
LSD	*The Logic of Scientific Discovery*, 1959, 1980[4]
MF	*The Myth of the Framework*, 1994
OK	*Objective Knowledge*, 1972, 1979[2]
OS	*The Open Society and Its Enemies*, 1945, 1966[5], 2 vols. (*OS1*, *OS2*)
P	*Postscript to The Logic of Scientific Discovery*, 1982–1983, 3 vols. (*P1*, *P2*, *P3*)
PH	*The Poverty of Historicism*, 1957, 1976[9]
RMC	*"Replies to My Critics,"* 1974
SIB	*The Self and Its Brain*, 1977, 1981[2]
SBW	*In Search of a Better World*, 1984/1994
TWP	*The World of Parmenides*, 1998
UQ	*Unended Quest*, 1974, 1976[2]
WP	*A World of Propensities*, 1990

Note

References to items from The Karl Popper Archive (Hoover Institution on War, Revolution and Peace, Stanford University) are in the form (x.y), where x is the number of the box, y the number of the folder; all other references are to the bibliography at the end of the book. All known English translations of foreign works I refer to are listed in the bibliography; all English translations not explicitly mentioned there are mine. All references in the text are by abbreviation, in case of Popper's own works, or else in round brackets and to the first edition of the work cited (if this is not available, square brackets are used instead); if required, a comma separates the first edition from the one actually used, whilst a slash indicates a translation.

Previous versions of the material presented here appeared, in various forms, in the following publications of mine: "The Ethical Nature of Karl Popper's Solution to the Problem of Rationality," *Philosophy of the Social Sciences* 32 (2002): 240–266; "The Positive Power of Negative Thinking," *Cladistics* 18 (2002): 446–452; "A Plea for Criticism in Matters Epistemological," *Social Epistemology* 17 (2003): 161–168; "Karl Popper's Philosophical Breakthrough," *Philosophy of Science* 71 (2004): 448–466; "Due approcci al problema della razionalità," *in Esperienza e razionalità: Prospettive contemporanee*, ed. Roberta Corvi, 60–77 (Milan: Franco Angeli Editore, 2005); "Rationality without Foundations," in *Karl Popper: A Centenary Assessment*, vol. 2, *Metaphysics and Epistemology*, ed. Ian C. Jarvie, Karl Milford, and David W. Miller, 131–144 (Aldershot: Ashgate Publishing Limited, 2006).

Acknowledgments

This book is the outcome of some twenty years of study, research, and discussion on the philosophy of Karl Popper. As for other works, I owe more than I can express in words to many friends and colleagues for their help, support, and encouragement. First and foremost to Karl Popper himself—Sir Karl, as he kindly allowed me to call him—who not only provided the ideas that shape this work, but also encouraged me, with unusual kindness and generosity, to criticize them and see them as the possible starting point of my own philosophical development. While writing these pages, the memory of our long conversations and walks in Kenley came to my mind stronger than ever. Particular gratitude I owe to Nimrod Bar-Am, Pierluigi Barrotta, Roberta Corvi, Malachi Hacohen, Mark Notturno, John Preston, Ferdinando Vidoni, John Wettersten, and especially Joseph Agassi, who never tired of discussing with me the various aspects of this work: though the responsibility for what follows is obviously only mine, they improved on it greatly. Finally, my heartfelt thanks to my parents, who made everything possible.

Introduction
Critical Rationalism

One of the most significant thinkers of the twentieth century, Karl Popper centred his whole philosophy around the intrinsically fallible character of our knowledge. According to him, the edifice of science does not rest upon solid bedrock, but plunges its roots into a muddy swamp:

> The empirical basis of objective science has [. . .] nothing 'absolute' about it. Science does not rest upon solid bedrock. The bold structure of its theories rises, as it were, above a swamp. It is like a building erected on piles. The piles are driven down from above into the swamp, but not down to any natural or 'given' base; and if we stop driving the piles deeper, it is not because we have reached firm ground. We simply stop when we are satisfied that the piles are firm enough to carry the structure, at least for the time being.[1]

Across seven decades of philosophical consideration, Popper elaborated a view at the same time coherent, rigorous, and unsteady, that sees in the constant struggle with problems the meaning and goal of life itself. His philosophy provides a middle way between two opposed authoritarian approaches to science and society: dogmatism and relativism. It offers an account of how scientific knowledge can be objective and rational without being certain, and without appealing to induction or grounding itself upon expert opinion, consensus, and authority of any kind.

The core feature of Popper's thought—the key to understanding his ideas on objectivity and rationality, as well as on politics and society—is that knowledge is *not* a form of justified belief. The majority of philosophers have regarded it thus: in their eyes, knowledge is justified belief—it is objective and rational if and only if it can be justified—and an argument is a justification if and only if it is rational and objective. It was this idea that, throughout the centuries, gave rise to the great foundationalist programmes that, in order to avoid infinite regress, appealed each time to the authority of reason or that of experience: indeed, we might read the great part of Western philosophy as the story of the rebellion against one authority or another, of the clash between competing authorities.

However, such efforts proved to be in vain, betraying the impossibility of any attempt to replace one authority with another (from Aristotle to the Bible, from reason to experience): for each of them proved to be not only an inadequate justification, but also fallible and questionable in itself. In his magnum opus, *Logik der Forschung* (1935; later translated into English in 1959 as *The Logic of Scientific Discovery*) Popper exposes the errors of any attempt at providing a foundation for our knowledge and describes science as empirical but not inductive, testable and confirmable but never certain, demarcated from metaphysics by falsifiability without deeming metaphysics meaningless.

In sharp contrast with other epistemologists, particularly the logical positivists, Popper gave up the idea that justification is a necessary requirement for scientific knowledge: our knowledge *cannot* and *need not* be justified.[2] Critical rationalism—as Popper labels his own philosophy, characterizing it by the appeal to reason and the role of criticism[3]—draws attention to the relevance of attempts, as to the way in which knowledge grows, and of criticism, as to the way in which it is put to test. Popper himself described this process by saying that knowledge progresses by conjectures and refutations, by bold attempts at solving problems checked by thorough and uncompromising tests. No room is left, within critical rationalism, for what the epistemological tradition has deemed its central question: whether, and to what extent, our knowledge has certain foundations and, in that case, what kind of foundations they are.

There is no method of discovering true theories (a recurrent illusion in Western philosophy: Plato, Aristotle, Francis Bacon, René Descartes, and John Stuart Mill, to mention but a few), nor—a weakened version of this illusion—can we ascertain the truth of a scientific hypothesis: we can never verify it. Neither (a still weaker version) can we ascertain whether a hypothesis is probable, or probably true.[4] Nevertheless, our knowledge is in a way "objective," since it can provide proofs of the falsity of a theory, and means by which we can learn from our errors.

Growth of knowledge and criticism are closely interconnected: according to Popper, we should prefer the theory which, at any given stage of critical discussion, accomplishes a growth of the possibly corroborated empirical content (that is, which has survived sincere attempts of refutation.) There is no inductive process through which theories can be confirmed: within Popper's philosophy of science there is no place for any theory of justification. His anti-inductivism exposes, in the first place, the myth of foundationalism and of the first (or ultimate) elements on the basis of which we can allegedly construct, or reconstruct, the world. On the other hand, pride of place is given to metaphysics, which Popper, opposing the logical positivists, refused to reject as meaningless, and rehabilitated as part and parcel of scientific research. It does not matter whether metaphysics is not empirically testable: any theory must be taken into consideration as far as a theory can be rationally criticized. In other words, we have to look for its fruitfulness, its ability to solve problems, to shed new light upon them and to set new ones.

As opposed to many philosophers who, confronting "Fries' trilemma" (infinite regress, dogmatism, or psychologism) either opted for psychologism or for some form of dogmatism, thus weakening their ideas of justification or truth, Popper gave up the idea that justification is a necessary requirement for scientific knowledge: this cannot and need not be justified. The process of testing our theories does not produce incontrovertible results, since these very results are nothing but hypotheses that need to be tested in their turn. It is a process without a natural end and that may, in principle, go on *ad infinitum*. Any decision to cut it short and to accept a statement is conventional: any statement always lacks a definitive verification, or a foundation, and may always be revised in the future.

Convention and experience modify, rather than determine each other. Corroboration is an assessment of how well a theory stands up to tests: it represents a rough estimate that has no implication for truth-value or probability. Our experiences are theory laden: theory informs all actions and decisions, and these cannot be justified. Experience and access to reality remain problematic, but it is possible to learn from it all the same. Metaphysical realism is a necessary working hypothesis, though there is no way of knowing for sure what the facts are and whether a statement actually corresponds to them: truth (*i.e.*, correspondence) remains an ideal, a regulative idea, always sought and never sure to be obtained (indeed, we may well obtain truth, but not justified certainty that we have obtained it).

Popper's critical rationalism is closely associated with the search for truth. As human beings, we should be aware of our fallibility and critical of our theories—but we can move from the awareness of our fallibility to the criticism of our theories only if we deliberately aim at the truth.[5] That is why truth plays, for Popper—as opposed as for Kuhn and Wittgenstein, for instance—the role of the regulative idea of scientific research and rational discussion. *Rationality requires no foundation, only critical dialogue*: this spells the end of foundationalist philosophy.

Whereas foundationalist philosophies equate the rationality of scientific knowledge with its justification and this, in turn, with logic (that is, with deductive or inductive argumentation), Popper established an equation among rationality, criticism, and logic, taking the latter to mean exclusively deductive argumentation. Indeed, valid deductive arguments are the only ones that allow us to transmit truth from premises to conclusion: for it is impossible—that is, inconsistent or contradictory—that a deductively valid argument has true premises and a false conclusion.[6]

We criticize a statement in order to show that some of its logical consequences are false. Criticism, that is, tries to show that a theory is false by showing that it is not coherent, either with itself or with other statements we deem true. However, with the single exception of contradiction, logic alone cannot prove the falsity of a statement. For if two statements contradict one another, logic can only conclude that at least one of them is false, but it does not—and cannot—indicate which one. And since no statement can be justified (or proved true), it follows that the acceptance or the rejection of criticism always involves a judgement.

In other words, *rationality is not so much a property of knowledge, as a task for humans.*[7] It is not the content of a theory, or a belief, that is rational, but rather the way we hold it (that is, the way we defend or attack it). Appealing to reason means nothing but taking a decision:[8] more than once Popper clearly acknowledged that his propensity for method as criticism (and, in his political philosophy, for the open society) is itself a choice. As such, it involves a moral element. However, the very decision in favour of rationalism cannot be rationally justified. This renders the balance achieved by Popper's philosophical edifice unsteady—but instability is the fundamental feature of life. We are rational as long as we remain open to criticism and willing to change our views when faced with criticism we think valid. We are rational, that is, as long as we are willing to appeal to reason and argument—as opposed to violence and force—to settle our disputes.

Reason, in Popper's eyes, is the negative faculty of relentless criticism. In Joseph Agassi's words:

> the whole importance of the refutation of received opinion, or of the best scientific opinion or idea or proposal, is just this: refutation opens the road to innovation. Nothing is more potent heuristic than refutation. Nothing is more conducive to progress than criticism of the current situation, nothing more likely to herald the new than discontent with the old. Criticism is liberation. The positive power of negative thinking.[9]

Criticism, freedom, and rationality also constitute the core of Popper's view of politics and the open society. For our actions may have unintended consequences: this follows from the rejection of any form of justificationism. This is particularly true when we seek large-scale political changes. As a consequence, we should not run the risk to make irrevocable and uncontrollable mistakes.

Popper prescribes a view of science *in itinere*, for it is easy to get rid of its products (of new and imaginative theories, that is) if they prove to be wrong. Along the same line, he prescribes a piecemeal reformist activity for society: for the consequences of our actions, which are already difficult to foresee, are nearly always impossible to control. Society and politics do not primarily derive from science their choices or means, but above all its method. Scientific rationality and democratic government are one and the same thing: for despite its defects, only democracy provides an institutional structure that allows for the use of reason in the political arena.

An outline of Popper's political philosophy, however brief, is beyond the scope of the present work, which is concerned with Popper's philosophy of science. It is important to realize, though, how his works in this field are intimately connected to his anti-justificationism, as it emerged from the intense years that culminated in Popper's philosophical breakthrough and, eventually, in *Logik der Forschung*.[10] Both *The Poverty of Historicism* and *The Open Society and Its Enemies* grew out of the theory of knowledge developed in *Logik der Forschung*, as well as of Popper's belief that our (mostly unconscious) opinions about the theory of knowledge and its central problems are decisive for our very attitude towards society and politics:

> rationalism is an attitude of readiness to listen to critical arguments and to learn from experience. It is fundamentally an attitude of admitting that "*I may be wrong and you may be right, and by an effort, we may get nearer to the truth*" [. . .] In short, the rationalist attitude, or, as I may perhaps label it, "the attitude of reasonableness," is very similar to the scientific attitude, to the belief that in the search for truth we need co-operation, and that, with the help of argument, we can in time attain something like objectivity.[11]

From the early 1930s, when he was writing *Die beiden Grundprobleme der Erkenntnistheorie* and *Logik der Forschung*, Popper openly confronted Kant's intellectual legacy and deemed himself as an unorthodox Kantian, for two reasons. First, his fallibilism: as opposed to Kant, Popper's is a critique of *uncertain* reason, in the strong sense of the word; for not only has science itself left behind the "idol" of certainty, but philosophy of science must also give it up. Second, his pluralism: what many regard as one of the worst flaws of the Western world, the impossibility of tracing it back to any unitary "principle," is for Popper (as opposed to Kant, who deemed Newton's theory as the true description of the world) its best virtue. As a scientific community, we should be proud not to have one single idea, but many, good or bad; and, as a society, not to have one single faith, or religion, but many, good or bad.

Certainty is perhaps one of our fundamental needs, but uncertainty characterizes our human condition, always accompanying even our best results.[12] We have the right to nourish such a need, but at the same time we have the duty to acknowledge our limits. Whereas the conceit of our reason paves the road to serfdom, the awareness of our ignorance is the very basis of our freedom: it is the key element of an epistemologically informed anthropology. As we read at the end of *The Logic of Scientific Discovery*, in one of the most beautiful philosophical passages of the twentieth century:

> With the idol of certainty (including that of degrees of imperfect certainty or probability) there falls one of the defences of obscurantism

which bar the way of scientific advance. For the worship of this idol hampers not only the boldness of our questions, but also the rigour and the integrity of our tests. The wrong view of science betrays itself in the craving to be right; for it is not his *possession* of knowledge, of irrefutable truth, that makes the man of science, but his persistent and recklessly critical *quest* for truth.[13]

1 Young Popper's Intellectual Revolution

In the early 1930s, Popper's *Logik der Forschung* (actually published in November 1934) came to solve two traditional philosophical problems: the problem of demarcation between science and non-science (by what criterion do we decide which hypothesis is scientific?) and the problem of induction (which has many formulations and variants, among which Popper chose the following: how do we learn from experience?). The traditional answer to the first question is that all and only established theories belong to science, while the traditional answer to the second is that experience leads to the adherence to established theories.

Popper's answers were different. First, he said that those hypotheses are scientific which are capable of being tested experimentally, where tests of a hypothesis comprise attempts to refute it. Second, he argued that learning from experience is the very act of overthrowing a theory with the help of that experience: we learn from experience by repeatedly positing explanatory hypotheses and refuting them experimentally, thus approximating the truth by stages.

This conclusion took a long elaboration, which is very interesting both from the historical and the philosophical point of view. Historically, it is interesting to see *how* Popper came to his original solutions, while philosophically it is interesting to understand *why* he changed his views.

AUTOBIOGRAPHY AND PHILOSOPHY

The best way to approach and understand Popper's thought is undoubtedly the historical one. In his autobiography Popper offered a picture of his own intellectual development. Such a reconstruction, however, is not always reliable and in this case tends to obscure, rather than illuminate, the intellectual revolution he underwent as a young intellectual. As an intellectual autobiography, *Unended Quest* presents the path Popper followed from one philosophical problem to another, leaving little place for personal elements or the context in which his ideas grew. By imposing his own mature views upon his memories, Popper portrayed his intellectual evolution as a sort of linear progress: from the youthful involvement with politics, to psychology in the 1920s, to the logic of science in the 1930s, to the work on the open society during the Second World War, until the "metaphysical" developments of the early 1950s.[1] Deeming science an adventurous revolutionary endeavour, Popper wrote his own autobiography so that a sort of rationality of scientific revolutions underlies the narrative, concealing the plurality of directions in which he moved, the different options he had to face, the intellectual *impasses* and the crucial turns.[2] The publication of Popper's early German writings—a few articles and three important theses, which were made available to Popper scholars in edited form only in 2006—allows now for a completely new picture.

In the 1920s Austria witnessed the rise to prominence of a vast, socialist-informed movement for the reform of primary and secondary school.[3] Its undisputed leader was Otto Glöckel (1874–1935), author of *Die österreichische Schulreform* (1923) and *Drillschule, Lernschule, Arbeitschule* (1928). In order to promote school reform, the Vienna city council merged the pedagogic and psychological institutes, thus establishing a two-year teacher training programme that combined academic and practical training. Among the teachers were Moritz Schlick (1882–1936), professor of philosophy of science, and Karl Bühler (1879–1964), theorist of the school-reform movement, a pupil of Oswald Külpe (1862–1915) and a member of the so-called Würzburg School of psychology.[4] The Pedagogic Institute allowed students the possibility to attend lectures at the University of Vienna, and Popper took this opportunity, particularly to attend lectures by the mathematicians Wilhelm Wirtinger, Philipp Furtwängler, Eduard Helly, Kurt Rudmeister and, above all, Hans Hahn, from whom he learned to see the history of thought as a succession of problems.

At the time he entered the Institute, Popper was a high school student with leftist political views, looking for a future. The Institute provided him with the opportunity to carry out a socially valuable job, that of teaching the lower classes. He thus set up studying pedagogy, but was also invited to attend lectures in psychology, philosophy, and sciences, as well as cultivating his love for music. The works written during the years spent at the Pedagogic Institute (1925–1927)[5] are crucial if we want to follow and attempt to reconstruct the different phases of the complex process that turned Popper, within the span of eight years, from an aspiring school-teacher to a mature philosopher.

Popper's very first publication dates from 1925: "Über die Stellung des Lehrers zu Schule und Schüler: Gesellschaftliche oder individualistische Erziehung?"[6] appeared in *Schulreform*—the official organ of the school-reform movement, together with *Die Quelle*.[7] It deals with the attitude of teachers towards their pupils, and argues that pupils should be regarded as much as possible as individuals, rather than as kinds. Popper would stick to this idea for the rest of his life, but this brief article plays no significant role in his early intellectual development. By contrast, the years between 1927 and 1931 are of the highest relevance. For the five essays written in this period turn out to be crucial: they comprise two longer articles, published in *Die Quelle*, and above all three theses, the most important of which (the ones completed in 1927 and 1929) only became available to scholars after Popper's death, in 1994.

FROM *HEIMAT* TO RATIONALITY

The first of the mentioned essays, "Zur Philosophie des Heimatgedankens" (1927),[8] deals with the problem of how to make students abandon the restricted perspective acquired in the environment in which they were born, in order to embrace wider views about culture, law, and rationality. Written after a seminar held at the Pedagogic Institute, it argues that learning starts from beliefs and habits that each individual learns within his *Heimat*, that is, the environment in which each individual was born and with which he identifies himself. Teachers should not overlook the cultural background of their pupils, and should rather take it as the first, necessary step towards their future intellectual development. However, the teachers' task is to help pupils gradually free themselves from it, in favour of wider and richer perspectives. As Popper would later argue, each person has the natural propensity to develop some kind of dogmatism that should later be abandoned through the conscious appeal to reason.

The pedagogy of the school-reform movement highlighted how learning should be rooted in the environment in which children grow up, that is, in their *Heimat*. The word *Heimat*, however, was too much laden with political meanings, also referring to a strong sense of belonging to one's homeland, both at the local and at the national level. The organizers of the seminar—led by Eduard Bürger, editor of *Die Quelle*—were trying to combine *Heimat* and socialism, and the school-reform movement aimed at forming the citizens of a modern nation. Together with other members of the *Verein freie Schule*, Glöckel was close to the *Deutscher Schulverein*, a group upholding nationalist and anti-Semitic views. Popper (whose family was of Jewish descent) was very far from all that, and in his 1927 article he criticizes the educational limits and political dangers of the concept of *Heimat*, in favour of designing a socialist and cosmopolitan view for progressive education.

Upon completing his coursework at the Pedagogic Institute, Popper submitted a protothesis, his *Hausarbeit*, carried out a training period in a secondary school, and took the final exam. The thesis, *"Gewohnheit" und "Gesetzerlebnis" in der Erziehung: Eine pädagogisch- strukturpsychologische Monographie* (1927),[9] is an attempt to provide a psychological explanation for children's innate need of dogmatism.

Popper asks whether psychology can be scientific. He is sure the question can be answered in the affirmative, but notices that nobody has ever provided an adequate explanation of what a scientific psychology would be. He therefore sets himself this task. In order to provide such an explanation, he offers a theory which he himself admits is unoriginal. Indeed, although he never says so explicitly, Popper clearly adopts a position very close to Hans Vaihinger's:[10] in order to be scientific, psychology must be able to decide whether a statement is true or false on the basis of the available empirical evidence. This is a necessary prerequisite for any discipline aspiring to achieve the status of science. Therefore, the aim of the philosophy of science becomes that of explaining how scientists assign truth-values to statements.

This is Popper's first attempt to develop a philosophy of science:[11] it is a philosophy of science with a strong inductivist flavour, despite its deductivist methodology. For, on the one hand, Vaihinger's theory is deductive: it states that theory must necessarily open the way to any empirical research, by pointing to its object. On the other hand, however, Popper wants to find a way to make theory guide research without allowing it to play a decisive role with experience. In order to avoid any circularity, he clearly states that the study of phenomena must precede the theoretical phase, and that facts should be observed in a purely neutral way, independently of any theoretical prejudice. The neutrality of observation and its primacy over theory are characteristic features of an inductivist approach—an approach that in 1927 Popper explicitly adopted but that he would relinquish later on.[12]

In fact, Popper opens his 1927 thesis as follows: "The present study, although highly theoretical in its main parts, has entirely arisen out of practical experiences [with pupils], and has eventually to serve practice again. Its method is therefore essentially inductive."[13] As opposed to Freud's, Adler's, and others' psychological theories, that often go beyond what is factually verifiable and even impose their points of view upon the facts, the theories of natural science, he says, only abstract from empirical data, never asserting something beyond the facts.[14]

The inductive process follows three steps: "an unprejudiced description (phenomenology) of empirical facts,"[15] that is, a comprehensive description of the psychology of "lawfulness" among children; an "attempt at abstraction";[16] and an "ordered representation (from a theoretical perspective),"[17] so as to offer an explanation of the phenomena taken into consideration. Interestingly, Popper also highlights an intrinsic danger in the inductive method, one that particularly threatens psychoanalysis. For, the phenomenology (the first step of the inductive process) always receives feedback from the theory governing the research (third step), so that the facts are selected in order to fill certain gaps in the theory:

> As time goes by, systematics penetrates deeper and deeper into phenomenology; this entails the danger that phenomenology—the description, which must be purely empirical and unprejudiced—gets influenced by

the theory itself. As a consequence, it [phenomenology] presupposes the theory for which it should provide the inductively based empirical evidence in the first place; this would clearly be a so-called circular reasoning, a *petitio principii*.[18]

However, he is not interested in solving this problem, for he thinks that a solution was provided already by Vaihinger. Scientific research requires theories that demarcate the range of facts that are to be studied; as such, however, theories should be deemed "fictions": they are mere organizational tools, not true (or false) descriptions of facts. Once a particular fact to be studied is identified, we must observe it in a neutral, objective way: only then do "fictions" acquire the status of scientific theories, or hypotheses (in Vaihinger's terminology).[19] If, then, as a consequence of a testing process, they get verified, we can regard them as theories, on an equal level with those in the natural sciences.[20] However, in order to be empirical, it is crucial that a theory "may only be formed inductively, through abstraction from empirical facts, and may never be projected into them."[21] Before getting to this stage, we need "fictions."

On the basis of this theory Popper deals with the problem of developing an adequate theory of critical thinking. Karl Bühler clearly distinguished between critical thinking and mechanical association, but an equally clear distinction between critical thinking and dogmatic thinking was still lacking. As opposed to dogmatic thinking, Popper says, critical thinking is described by the theory of judgements: we should doubt any statements that have not been adequately justified.[22] He had difficulties overcoming the inner tension in his own view between the idea that thinking played an active role and the inductivist idea that regarded judgement as something passive, only accepting what is justified, rather than actively looking for errors. Consider the following passages:

> Free thinking [. . .] is critical thinking—dogmatic thinking is not free: under free thought one can only understand thinking "without prejudices," that is, thinking which judges states of affairs, without presupposing the result (the judgement) of judging; critical thinking is also thinking with foundations; it is active, spontaneous and autonomous thinking, in contrast with dogmatic thinking, which does not touch the accepted (adopted) "judgement."[23]

> We shall use the terms "dogmatic" and "critical" in a more liberal sense than is often usual: by *"dogmatic" thinking* we wish to mean a type of thinking that is characterized by the mere acceptance of and sticking to certain principles. These principles are *adopted*, "blindly" taken for true, without even recognizing the possibility of their falsehood; in particular, they are not investigated as to their correctness through experience; one *sticks* to them, that is, one applies them stubbornly wherever they seem to be applicable.[24]

Critical thinking, by contrast, may be characterized by such questions as "is it really true (or "correct")?, "does it really have to be that way?"; also in the field of ethics: "is it really good and right?", etc. Critical thought attempts to question the principles that are usually at first blindly received and dogmatically maintained, in order to accept and apply them first after *verification*, particularly by means of experience, but also through reflection."[25]

We can see in these passages elements that very well fit with the sceptical attitude and the critical view Popper would adopt a few years later. However, they are mingled with ideas that clearly betray an inductivist stance. These early methodological remarks do not display any feature of Popper's deductive method by trial and error. His stance is thoroughly traditional, that is, inductive—and the same can be said of his ideas concerning demarcation.[26] From the inductivist point of view, the problems of induction and demarcation are inextricably combined: only those theories that have been justified can be deemed scientific. It is the same view adopted in *Die beiden Grundprobleme der Erkenntnistheorie* (without §11): here Popper states that only decidable theories are scientific.[27]

POPPER'S DOCTORAL DISSERTATION AND THE CONFRONTATION WITH THE WÜRZBURG SCHOOL

Popper progressively moved from psychology to philosophy of science by discussing the methods of the psychology of thought. He had already begun to do so in the 1927 thesis, as we have just seen, but he did so more thoroughly and significantly in his doctoral dissertation, *Zur Methodenfrage der Denkpsychologie*, which was examined by Schlick and Bühler in 1928.[28]

In the dissertation Popper discusses the use of philosophical anthropology as a guide for research in psychology. He asks, in particular, whether it is to be assumed—as Schlick and the *Gestalt* psychologists had done—that psychological processes can and must be reduced to physical and/or biological processes. This is a new version of the methodological problem he dealt with in his 1927 *Hausarbeit*: how is it possible to employ a theory as a guide for research without compromising the way in which facts should be seen? In his doctoral dissertation Popper abandons the distinctions of Vaihinger's theory but follows the very same line of argument he had followed one year earlier: in order to guide research, we should appeal to theory, but we must not prejudicially commit ourselves on the basis of the data supporting it.[29]

Besides Schlick and Bühler, Popper also takes into consideration Else Köhler. Bühler had followed the non-reductionism of the Würzburg School, whereas Köhler's works defended a strong reductionism of psychological processes to the physical ones, as Schlick had suggested. Popper takes his stand on the Würzburg School's side, but his is a qualified adhesion: psychological processes, he contends, can be either reducible to physical ones or not, but it is important not to decide the issue *a priori*, as Köhler and Schlick (as opposed to Vaihinger) wanted to do.

In the first part of the dissertation Popper deals with the relationship of psychology to physics and biology, arguing for the idea that psychological research needs be kept independent from other areas of research. He does not deny that it might be reduced to other areas, such as physics (as Schlick suggested), though biology is a more likely candidate. But he insists that this issue cannot be decided within philosophy: the correct option must be a consequence of actual empirical research. Any *a priori* decision would be unacceptable, for it would go beyond the limits of empirical research.

The methodological separation between physics and psychology is a premise necessary for the discovery of any reduction of one aspect to another: only an open mind and empirical research can lead us to decide which is the correct hypothesis. If Schlick upheld physicalism as the only possible scientific approach, Popper deems it a hypothesis like many others that need be tested. In other words, his is not a mere defence of the inductivist approach, for such a perspective would not invite the consideration of a variety of alternatives before appealing to the facts, whilst Popper is clearly concerned with the consideration of a variety of theoretical alternatives before undertaking empirical research.

The critique of physicalism occupies over half of the dissertation. In the second half, Popper shows how cognitive psychology, as well as language theory, works at three levels: experience, behaviour, and thought structure. Following Bühler, he shows how each of these levels is necessary and requires a different methodology. The necessity of experience had been showed by the Würzburg School and, problems connected to the use of experience or the method of introspection in the study of thought notwithstanding, no cognitive psychology as such can do without it. Behaviour, on the other hand, is required in order to explain certain reactions observed in animals that appear at times to be oriented towards an aim. Finally, cognitive structures are indispensable to explain—for example—meaning: the study of the descriptive function of language, for instance, cannot do this by itself, and other factors need to be taken into consideration as well. The analysis of the descriptive function appears problematic, too, for it raises problems related to the relationships between theory of knowledge, logic, and psychology.[30] The development of a complete psychology cannot overlook the solution of these problems and must be integrated—Popper argues, thus following Bühler once again—with the consideration of biology.

This is a crucial analysis, since shortly after having completed his dissertation, Popper abandons psychology to devote himself to the solution of the problems discussed in the text. The aim was to separate psychology from methodology and to construct a deductivist theory of science on the basis of Selz's psychology. Popper himself suggests this move, while discussing the possibility that Selz's cognitive psychology might describe both scientific research and prescientific thought. Selz had employed thought psychology to analyze individual episodes in scientific research, but—according to Popper—he overlooked the stronger hypothesis that thought and scientific practice as a whole could be actually described in such terms. Popper writes:

> Are there maybe important parallels in the methodologies and operations between the scientific and the "prescientific" induction? Just to point out one example: the Selzian concept of "probing behaviour" seems to me to have obvious parallels in the objective practice of science. Science also probes its theories—its "models," as Bühler says— and, indeed, in a way which fully corresponds to the Selzian scheme. The actual paths of research in science do not correspond in any way whatsoever to logical principles of presentation, just as little as the "operations" described by Selz correspond to objective logical "operations." Nevertheless, in the long term the practice of science is clearly "steered by tasks"; the "determining tendencies" stand out clearly.[31] Selz himself numerously used the example of scientific research; admittedly not in the sense that we suggest here. It is not the objective mental "forms" and not objective scientific practice that his analysis covers, but certain individual scientific discoveries are analysed as an aspect of experience.[32]

At the time he received his doctorate, Popper had made the non-reductionism of the Würzburg School his own; he had tried to develop his own psychology that apparently fitted very well within the framework of the School;[33] and he had felt the urge, within Bühler's scheme, to reconsider the problems of the separation between logic, methodology, and psychology, as well as that of extending the study of thought beyond the descriptive function of language—that is, to integrate it with the psychology of thought, particularly that of the Würzburg School.[34] If we add to all that the need for a methodological theory that is able to replace Külpe's, we get a fairly precise picture of the condition in which Popper found himself at the time when he was about to begin his work on methodology.

Further than pedagogy, psychology, and the realism of the Würzburg School, however, Popper needed to address the philosophy of science prominent at that time, that of Logical Positivism. The construction of a deductivist psychology by Külpe, Bühler, and Selz had showed that a deductivist methodology was needed. Popper became aware of this need when he defended the psychology of the Würzburg School in the process of writing his dissertation. Particularly important proved to be the need to account for recent developments in logic: the Würzburg School had not done that, and this might be one of the main reasons why Popper progressively distanced himself from it.

As Popper learned from Schlick, the task of developing a new deductivist philosophy of science can be undertaken only if the new developments in logic are taken into account, as well as the ways in which they were employed within the logical positivists' epistemology.[35] The next step was therefore to study the philosophical stance of the members of the Vienna Circle. The problem immediately posed was how to reconcile Popper's by and large deductivist view with the logical positivists' inductivism. The first book of *Die beiden Grundprobleme der Erkenntnistheorie* (without §11) contains Popper's analysis of the problem, his second attempt to develop a philosophy of science.

THE 1929 THESIS ON GEOMETRY: POPPER'S FIRST STEPS TOWARDS A NEW PHILOSOPHY OF SCIENCE

Popper's first meeting with mathematics, logic, and the philosophy of the natural sciences took place in 1929, with his third thesis, *Axiome, Definitionen und Postulate der Geometrie*.[36] While its declared intent is that of gaining the qualification to teach mathematics and physics in secondary schools (*Hauptschule*), Popper's aim was actually to gain familiarity with and master these disciplines. The knowledge he acquired while writing the thesis will prove of crucial importance for the subsequent discussion of induction and will be evident both in *Die beiden Grundprobleme der Erkenntnistheorie* and *Logik der Forschung*.

The fracture with his previous studies is significant: if until 1929 Popper was concerned with epistemology as far as it dealt with the problem of the justification of the scientific status of psychology and pedagogy, in his *Habilitation* he explicitly discussed the cognitive status of geometry without making any reference to its psychological and pedagogical aspects. The 1929 thesis contains the first formulation of the problem of scientific rationality, enabling Popper's future progress. Furthermore, it reflects Popper's shift of interest from cognitive psychology to the logic and methodology of science.

The thesis deals with the problem of the impact of non-Euclidean geometries on two groups of problems in the philosophy of science: those concerning the foundations of geometry (questions having to do with the axiomatic development of geometrical theories, such as the problem of the independence and necessity of the axioms, of the completeness and consistency of axiomatic systems, as well as whether definitions should be explicit or implicit) and those concerning the truth and falsity of geometrical assumptions, including the problems of the applicability of the various geometries to the real world. The thesis culminates with the latter discussion, and this is undoubtedly the most important issue for the subsequent development of Popper's thought.

The thesis is divided into three parts. In the first, Popper introduces the two groups of problems that have been just outlined. In the second, he goes over the history of the discovery of non-Euclidean geometries, highlighting two aspects: their role in the formulation of the problems of independency and necessity of the axioms, as well as of completeness and consistency of axiomatic systems; and their influence on discussions about the applicability of geometry to the real world. Finally, the third part deals with the second group of problems in the philosophy of science, those concerning the validity (*Gültigkeit*) of geometrical theories.

For the first time, with the discovery of non-Euclidean geometries, theoretical physics faced a choice among competing geometries to describe physical space: all of them were internally consistent, offered alternative conceptions of space, and were unconditionally true (in a purely formal way). There arose the problem of their application to reality. Logic, as well as mathematics, is never involved in questions of conformity to reality, since the relationships it establishes among objects are conceptual, not real, the only thing at issue is its internal consistency. The same holds for pure geometry: all geometric theorems are analytic, derived from axioms. As long as the axioms are accepted, the theorems are valid *a priori*. Applied geometry, however, is different: it establishes real, not conceptual, relationships among objects. Its theorems are, at least in part, synthetic and valid *a posteriori*. Their proof requires observational verification. Therefore, applied geometry involves scientific procedure and raises epistemological as well as methodological issues.[37]

This is the premise of the most important part of the text: the discussion of whether it is possible to determine the truth or falsity of the various geometries as descriptions of the world. Part of the problem, Popper says, is simple: following Einstein, he says that pure geometry does not apply to the world. It does not mean to do so, nor does its development depend on whether it does so or not. As a matter of fact, however, we do apply geometries to the world. As to their applicability, Popper discusses three positions. The first is Jules Henri Poincaré's: the choice of what geometry to apply to the world is purely conventional. Among the consistent and sufficiently powerful geometries available to us, he said, we can choose the one we prefer: only, we should choose the most convenient metric.[38] The second is Hermann von Helmholtz's, according to which the geometry that applies to the world might well not be Euclidean (for instance, Popper mentions, Gauss and Riemann thought so): however, this is a matter that it is only up to experience to decide.[39] The third position is Albert Einstein's: as far as geometries are certain they do not apply to reality, while as far as they apply to reality they are only approximations of the truth.[40]

Popper rejects both Poincaré's conventionalism and Helmholtz's inductivism: scientists can prefer one geometry to another (Riemann's geometry of curved but not infinite space, for instance) because they think it better describes the world—but they cannot, as Helmholtz suggested, ascertain the truth of one geometry by appealing to facts. In other words, Popper uses Poincaré's proposal—we can choose the simplest geometry and use it by convention, without being able to declare for ourselves as to whether it actually describes the world as it is—to criticize Helmholtz's. But Poincaré's proposal is inadequate, too, for it cannot account for the fact that when we choose a geometry we might want to integrate it with a physical theory. As Helmholtz highlighted, the physics associated with one geometry may reveal much more complexity than that associated with another, and it is for this very reason that Einstein adopted Riemann's geometry, and not Euclid's, in his general theory of relativity. Neither theory is adequate, then: in physics, on the one hand, we cannot—and should not—appeal to mere conventions, since our aim is to describe and understand the world; on the other hand, empirical evidence alone cannot determine which of the various available geometries is correct. Popper's proposal is that we can choose, as the best approximation to reality, the geometry that, in association with a physical theory, involves the simplest combination.

Popper adopts the methodological proposal according to which scientific theories require the greatest economy in the use of hypotheses.[41] Our choice among competing geometrical metrics describing space can then be decided methodologically: this is Popper's first statement in the methodology of the natural sciences. The path he intends to follow is clear: even though he accepts a strong conventionalist component within his methodology, Popper chooses to let tests have the last word over our decisions. The acceptance of such results certainly remains a matter of convention, but our choice bears an epistemic value: for, in any such decisions, our will to achieve a description ever closer to reality (that is, to truth) must play a decisive role. From now on, the idea of truth—the regulative idea of truth, in the Kantian sense—will remain at the heart of Popper's philosophy. Together with the idea of simplicity and the metaphysical assumption of realism, it opens a new way, rigorously deductive, which is far distant from both Poincaré's conventionalism and Helmholtz's inductivism.[42]

Applied geometry provides the context for Popper's discussion of scientific rationality. In the following years, he would apply the hypothetic-deductive model to all natural sciences (ones which, as opposed to geometry, are not axiomatic):

> whenever we are doubtful whether or not our statements deal with the real world, we can decide it by asking ourselves whether or not we are ready to accept an empirical refutation. If we are determined, on principle, to defend our statements in the face of refutations [. . .], we are not speaking about reality. Only if we are ready to accept refutations do we speak about reality.[43]

Thus we can read, in the closing section of the 1929 thesis:

Only now, after having surveyed the different species of pure and applied geometries [. . .] is it possible to formulate a conclusive judgement about their kind of validity: are they valid *a priori* or *a posteriori*, in a rigorous or in an approximate way? It is necessary to clarify that the opposition between validity *a priori* and *a posteriori* opens an insurmountable abyss between two sciences.[44] For the statements of an *a priori* science hold exactly, as purely conceptual constructions, while all the statements of an *a posteriori* science can be regarded as *approximately valid* only on the basis of ascertainment by observation and experiment, on the basis of "empirical verification." *Only* pure geometries are valid *a priori*, as has been shown. *Pure geometry* and the various kinds of applications of geometry are therefore *completely different* sciences.[45]

These remarks retain a strong inductivist flavour. In fact, the 1929 thesis marked only Popper's first step towards the unqualified rejection of inductivism in favour of a consistent deductivism. Nevertheless, it is a crucial step, for it is the problematic relationship between geometrical and mathematical constructions and physical reality that triggers Popper's philosophical revolution. It will be on the basis of the step taken in 1929 that, a few years later, in *Logik der Forschung*, Popper will overcome too direct a notion of such a relationship, casting doubts on inductive inference, thus starting to conceive in a new (consistently non-inductivist) manner the problem of the relationship between axiomatic systems and reality, between theoretical and observation statements.

There is still a long way to go, though. What, in hindsight, might appear as a simple and straightforward move was not yet sufficiently clear in Popper's mind.[46] He could not quite yet integrate the conclusions he reached in 1929 with what he took to be the central task for the philosophy of science, that is, to explain how we can decide the truth or falsity of scientific statements on grounds of empirical evidence. Only after the failure of this approach will Popper attempt to develop a theory of scientific knowledge that would involve no proof.

THE ATTEMPT TO PROVIDE A NEOPOSITIVIST THEORY OF KNOWLEDGE: *DIE BEIDEN GRUNDPROBLEME DER ERKENNTNISTHEORIE*, VOL. I (WITHOUT §11)

In 1931 Popper published "Die Gedächtnispflege unter dem Gesichtspunkt der Selbsttätigkeit," in which he applies Selz's psychology to a defence of the teaching methods proposed by Glöckel and Bürger. Appearing once again in *Die Quelle*, this would be Popper's last contribution to the Austrian school-reform movement.[47]

At the same time he was working on this article, Popper began to write *Die beiden Grundprobleme der Erkenntnistheorie*,[48] in which he attempted to develop a philosophy of science adequate to the standards of the time—and particularly those set by logical positivists. Such an attempt is documented in the first volume of the book, without §11 (Chapter 5). It is a deductivist theory, constructed in analogy with Selz's deductivist psychology. If Einstein's and Poincaré's ideas had allowed him to reject inductivism, they had not explained its possibility away, and Selz's deductivist psychology perfectly fitted the task.

Although he explicitly aims at rejecting the logical positivists' inductivism and conventionalism in favour of a consistent deductivism, Popper himself declared his own attempt as closest, among available alternatives, to that of logical positivists.[49] He clearly adopts their language and style. His purpose is to do philosophy of science through the logical analysis of scientific arguments.[50] Taking that approach to its limits, he will expose the impossibility of developing an adequate theory of science within that framework.

Popper's problem was how to distinguish scientific statements from nonscientific ones. His idea was to demarcate scientific statements on the grounds of the logic that characterizes the arguments through which they are credited with truth-values. In his eyes, the aim of a theory of knowledge—at the beginning of the 1930s he still used the term *Erkenntnislehre*—was to show whether science is able to decide the truth or falsity of a statement or not and, when it can do that, how it manages to do so. He also thought that the decision as to the truth-value of a statement can only be taken when its truth or falsity is proved. As a consequence, Popper's theory is a deductivist theory in which it is taken for granted that the aim of science consists in proof.

This is a crucial difference between *Die beiden Grundprobleme der Erkenntnistheorie* (without §11) and *Logik der Forschung*. In the former, Popper understands the task of the philosophy of science basically along the same lines on which he understood it in *"Gewohnheit" und "Gesetzerlebnis" in der Erziehung*, that is, to account for the way in which scientists determine the truth-value of statements.[51] By contrast, in *Logik der Forschung* the task of the philosophy of science becomes that of describing the methodological rules followed by scientists in their attempt to attain the truth. Between 1927 and 1933 Popper did not change his mind about this, the progresses achieved in his 1929 *Habilitation* notwithstanding—further evidence that in his third thesis we might only identify a few scattered elements of the point of view Popper would make his own in 1934, and that had a completely different weight just a few years earlier.[52] As a consequence of this way of understanding the task of the philosophy of science, in *Die beiden Grundprobleme der Erkenntnistheorie* (without §11) Popper does not even try to account for the growth of knowledge: he is concerned only with the logic of science, that is, how to explain the truth-values scientists assign to statements. Whereas for Reichenbach it is impossible to assign truth-values to scientific statements, but only probabilities,[53] Popper declares it is possible to assign the value "false" to some statements, but it is impossible to assign either the value "true" or any degrees of probability to universal statements. At the time he was working on *Die beiden Grundprobleme der Erkenntnistheorie* Popper was not interested in the rules of science but was concerned only with how to determine the validity of scientific statements. His criticism of Reichenbach's proposal, according to which scientific statements could be assigned a probability, will constitute the starting point of the theory expounded in *Logik der Forschung*—but it is not possible to trace such a theory back to the first volume of *Die beiden Grundprobleme der Erkenntnistheorie*, without §11, as John Wettersten has shown.[54]

The fact that he shared with Reichenbach the idea that this should be the task of the philosophy of science means that, at the time, Popper was not yet clear about the terms of the problem of demarcation—although, just like the logical positivists, he believed a way to distinguish scientific statements from metaphysical ones was required. As it is evident from the closing paragraphs of the text,[55] he was only aware of the fact that if statements are assigned a truth-value in the only way he then thought possible—that is, only the value "false," never the value "true" or the mark "probable"—then there arises a new problem of demarcation, different from the one he described at the beginning of the volume.[56] For, if it were indeed possible to ascertain the truth, or the probability, of statements, science would become a classification of such statements: it would become, as Popper declares at the very beginning, a problem of justification, or validity of statements. According to the view developed in the text, though, the problem is that of distinguishing scientific from nonscientific theories when no theory can be in any way proved.

Popper shared the view expounded in *Die beiden Grundprobleme der Erkenntnistheorie* (without §11) with all the philosophers of science who were concerned with demarcation at that time: all of them, Popper included, took for granted that any solution of the problem of induction would have led to a demarcation of science, understood as the set of statements of which it is possible to prove the truth, the falsity, or some degree of probability. If, however (as Popper himself realized while writing the book), it is not possible to prove the truth of a theory, such an assumption would have taken one nowhere. Therefore, he tried to say something more about falsifiability, but the very structure of the work did not allow him to reach a solution: for, thus stated, the only problem he faced was to explain how it was legitimate (or justified) to assign truth-values to statements. He did not formulate the problem of demarcation, yet: he merely realized that, since it is impossible to prove scientific theories, the problem must be restated and treated separately from the problem of induction.

Thus set, the first volume of *Die beiden Grundprobleme der Erkenntnistheorie* is an analysis whose aim is to establish whether and how statements can be proved true or false, and in a discussion of the problem whether scientific statements (which are universal, by their own very nature) should be actually regarded as statements, that is, whether they can be assigned a truth-value or not. Popper's conclusion is that science can show, by way of deductive arguments (the *modus tollens*), that some universal statements are false. It can do that because the falsity of some statements can be directly ascertained: if a singular statement is shown to validly follow from a universal statement, and we ascertain that such a singular statement is false, we can deductively conclude that the universal statement is false.

There is no method of showing that a universal statement is true, though—or, to put it differently: there is no inductively valid argument. Universal statements must be deemed "fictions" (in Vaihinger's sense):[57] they are not proper statements, since we can never ascertain their truth, but are very close to being so, since we can ask whether they are true or false.[58]

In the process of writing *Die beiden Grundprobleme der Erkenntnistheorie* Popper hoped to show that his theory of statements is a solution of the problem of induction. At the end of the text—at the end, that is, of what is presented as the first volume of the book—he asks whether he solved the problem of induction.[59] And he suggests—reminding the reader that according to Wittgenstein the problem of demarcation was the only fundamental problem of the theory of knowledge[60]—that there exists, in fact, one single problem, that of demarcation. And that a way to check whether he has actually solved the problem of induction is to see whether his proposal actually solved the problem of demarcation as well.

At this point, the discussion breaks down. Popper does not know how to continue. The discussion collected in what the editor deemed the remaining sections of the second volume does not provide any answer.[61]

Logical positivists had tried to demarcate meaningful statements, that is, empirical statements, by regarding them, and them only, as part of science. However, from the very beginning, Popper tried to avoid appealing to meaning as a criterion for ascertaining the scientific character of statements. At the time, though, no criterion for distinguishing scientific from nonscientific statements among nonverifiable ones was available to him.

The solution he would later advance, that is, scientific statements—he would then call them theories—are the refutable ones, was not yet available. Nor could it be: in order to establish which statements (or theories) are refutable, logic and methodology must be integrated. For it is not possible to distinguish refutable and nonrefutable universal statements as such, without appealing to methodological considerations. At the time, however, Popper stated the problem of demarcation in such a way as to bar any methodological considerations. Only later, when he would no longer speak of universal statements as such, but of explanatory theories and methodology, will he be able to declare that some theories are refutable and can be distinguished from nonrefutable ones, since they allow to deduce "basic statements" and are not defended by appealing to *ad hoc* stratagems.

Finally, it is interesting to note three key aspects of Popper's position, as described in the first volume of *Die beiden Grundprobleme der*

Erkenntnistheorie (without §11). The first concerns universal statements. In the whole text Popper regards them as "fictions" *à la* Vaihinger. He does so because he believes it is possible to assign truth-values only to statements that are proved right or wrong, whereas universal statements are not of this kind. He would share this (wrong) view with the logical positivists until his meeting with Tarski in 1935. For, according to the Polish logician, it is possible to assign truth-values also to statements that have not yet been proved true or false: by his definition of truth, in other words, there are true and as yet unproved statements.[62] This disavows Popper's belief that only proved statements can be assigned a truth-value— a belief that he maintained in *Logik der Forschung* and that prevented him from developing his view of science as conjectures and refutations.[63] Popper realized the problem, but avoided it in *Logik der Forschung* and confined himself to declaring that theories—he no longer refers to them as universal statements—can be interpreted from a realist stance.[64] And yet, he titled the book "logic of research," to answer Reichenbach's statement that there is no logic of science, that is, no proofs nor refutations.[65]

The second aspect refers to singular statements: in *Die beiden Grundprobleme der Erkenntnistheorie* (without §11) Popper says they are veridical; in §11, while discussing Fries' and Nelson's thesis, he begins to distance himself from this view; finally, in *Logik der Forschung*, "basic statements"—no longer "singular statements"—are always provisional and subject to methodological rules.

The third and final aspect concerns conventionalism: Popper had already rejected conventionalism in his 1929 thesis, but regarded it as a theory about the status of statements and criticized it as such, without appealing to methodological considerations. However, when Reichenbach replied to Popper's note, in the 1933 issue of *Erkenntnis*, saying that his proposal to demarcate systems of statements on the basis of their falsifiability had already been tried out and rejected by logical positivists, since any falsification could have been avoided with the help of *ad hoc* moves,[66] Popper began to look at conventionalism as a theory that poses a methodological problem—and introduced methodological rules in order to solve it.

THE METHODOLOGICAL TURN: *DIE BEIDEN GRUNDPROBLEME DER ERKENNTNISTHEORIE*, VOL. I, §11, AND VOL. II

At a closer look, §11 of the first volume of *Die beiden Grundprobleme der Erkenntnistheorie* shows some incongruities and inconsistencies with the rest of the book. It is the longest section; it appears to be independent of the remaining argument and it connects with it abruptly, unnaturally; it employs a terminology ("empirical basis," "basic statement," "observation statement") that is standard in *Logik der Forschung*, but lacking in the rest of *Die beiden Grundprobleme der Erkenntnistheorie*; finally, its antifoundationalism sharply contrasts with the other sections.[67] If there existed—as the editor Troels Eggers Hansen has argued—a preceding version of §11, it is plausible to think that the present version was written after it and replaced it at the last moment.[68] At any rate, together with the fragments that are now offered as the second volume of the book, §11 presents a different position from the one expounded in the other sections.

The central problem dealt with in §11 is foundationalism. Beginning in the early 1920s Popper discussed at length, with his friend Julius Kraft (1898–1960, a pupil of Leonard Nelson's) Kant's epistemology, and particularly the criticism raised by Jakob F. Fries.[69] Contrary to Kraft, who accepted Fries' criticism and his proposal for an alternative foundation for knowledge, Popper immediately rejected this solution as psychologistic. Then in the early 1930s, he gave up foundationalism completely.

In Malachi Hacohen's words:

> Kant held that certain propositions had an a priori validity because no conception of reality or morality was possible without them. Fries thought that these synthetic a priori propositions left too much of the world closed to the human mind and, at the same time, ran the risk of subjectivism. He developed a methodological procedure for grounding knowledge in universal human psychology, thereby eliminating much of Kant's agnosticism and "subjectivism." In his dissertation, Nelson defended Fries against contemporary Neo-Kantians. [. . .] His voluminous work in epistemology, ethics, and jurisprudence carried the imprint of Fries's "Kantianism with a greater confidence of reason." Popper rejected precisely this "confidence." He shared Fries's and Nelson's critique of Kant but declined their solution, and offered his own: ever uncertain knowledge. His arguments with Kraft over Fries and Nelson set the context for his epistemological revolution.[70]

According to Fries, the guarantee of the truth of the results produced by the activity of the psyche is not to be sought for outside the psyche itself, but must be rather recognized in what he called the "self-confidence" in reason, the only possible foundation for the objective validity of human knowledge. Nelson believed that epistemology was unable to solve the problem of knowledge. He opted for moving into the field of psychology: Fries' "immediate knowledge" was the only possible foundation for knowledge.

Popper accepted Nelson's formulation but not his conclusion. Fries and Nelson were wrong to rely on psychology: immediate knowledge is subjective and provides no firm foundation for science; it is expressed in statements, and these require a justification, thus leading to infinite regress. Fries' immediate knowledge is intuitive, dogmatic, and irrational. Fries, Carnap, and Neurath make the same mistake: in fact, protocols are nothing but psychological reports in disguise.[71] Experience, or the language that describes it, cannot exert direct control over science: science has no need for an empirical basis.

Popper's proposal is that scientific theories can never be conclusively verified, or declared true; they are regulative ideas, logical "fictions" constructed with the heuristic aim of deducing predictions and testing them. If Fries and Nelson had declared epistemology impossible due to the infinite regress in which any attempt to ground knowledge through inductive procedures is bound to fall, Popper, by contrast, states that what renders epistemology impossible is not infinite regress, but Fries' and Nelson's very foundationalism. Epistemology's task is not the search for a foundation for knowledge, but to provide general methodological rules for scientific research. It does not justify statements—rather, it indicates methods and criticizes procedures, highlighting contradictions and mistakes. Its aim is to elucidate, criticize, and improve scientific practice, not to provide a foundation for it.

Popper is now able to restate the central problem of epistemology. At its root, he says, lies "Fries' trilemma." For the attempt to provide knowledge with a firm foundation can end up either with the acceptance of basic statements which, we believe, are not in need of further justification (dogmatism), or by accepting infinite regress, according to which we recognize that no statement can ever be conclusively validated; or else—and this is Fries' and Nelson's option, as well as most philosophers', with the exception of conventionalists—with psychologism, that is, with the justification of statements by appealing to experience, to perceptions, and to "immediate knowledge." Popper disagreed: observation and experience reports are scientifically acceptable only if they can be intersubjectively tested. Scientists' own convictions, however important, are genetic-historical in character, not epistemological: they can contribute to the discovery of a theory or can help explain subjective preference for it, but cannot justify it. As opposed to many philosophers who, facing "Fries' trilemma" either opted for psychologism or for some sort of dogmatism, thus weakening their concepts of justification or of truth, Popper gave up the idea that justification is a necessary condition for scientific research: this cannot be and has no need to be justified. This spells the end of any foundationalist philosophy.

If Popper had earlier thought that infinite regress ended either with the verification or the falsification of particular predictions, he now stated that such predictions—in §11, as we have seen, he started to call them basic statements[72]—are theories with a lesser degree of universality: their test does not always lead to incontrovertible results. They are themselves new hypotheses, susceptible to further revisions. The testing process has no natural end and could, in principle, go on forever. Any decision to break it up and accept a statement is conventional: such a statement always lacks a conclusive verification—that is, a foundation—and can always be falsified and revised in the future. The acceptance of any such statement is dogmatic, in a sense, but it is a harmless form of dogmatism: as soon as new doubts emerge, scientists resume testing. Subjective convictions contribute to forge the consensus that leads to the suspension of the testing process but, again, it is an innocuous form of psychologism: scientists do not found their theories, and although they accept them on the basis of a methodological rule, agreeing to accept confirmed statements as scientific, no rule is so rigid as to force univocal moves in any situation. Dogmatism, infinite regress, and psychologism all play a role in scientific practice, but are rendered inoffensive by the hypothetic and falsifiable character of science.[73]

Popper kept in constant dialogue with logical positivists, and it was this confrontation that fuelled the revolution triggered by the solution of "Fries' trilemma" and the consequent relinquishment of foundationalism. Indeed, his previous attempt to elaborate a positivist theory of science generated a first objection by one of the leading members of the Vienna Circle: singular statements are not reliable, argued Neurath, since it is impossible to directly ascertain their truth or falsity. They must be deemed provisional, just as much as other statements. There can be no proof of the falsity of universal statements.[74]

The second objection came from Reichenbach. Popper had published a brief note in *Erkenntnis*—the official journal of the Circle—in which he outlined a view in between the one described in *Die beiden Grundprobleme der Erkenntnistheorie* (vol. 1, without §11) and the one expounded in *Logik der Forschung*:[75] He suggested demarcating scientific theories on the basis of their refutability, but did not explain the methodological rule that was required in order to apply such a criterion. Reichenbach objected that it was always possible to protect a theory from refutation by appealing to a small modification (*ad hoc* hypothesis): logical positivists[76] had long before considered Popper's proposal, but eventually rejected it for this very reason.[77]

Popper took up the challenge: the core of his solution is the revision of the two fundamental problems of the theory of knowledge that shifted from *Erkenntnislehre* to become philosophy of science. He realized that the objections raised by his critics could not be answered by appealing to the logic of scientific proofs, as he had tried to do thus far. He then gave up the idea of studying the way in which it is possible to prove the truth or falsity of statements, and introduced a set of methodological rules for science. He is now concerned with demarcation—not of statements as belonging to science, if proved true, or as not belonging to science, if falsified, though, but of theories subjected to the proper methods of scientific research.

In reply to Reichenbach, Popper advanced a methodological rule so as to prevent the protection of a theory from refutation as a consequence of small modifications (as highlighted by conventionalists). In reply to Neurath, he stated that it is not necessary to assume that the truth or falsity of singular statements can be determined directly and with certainty: each of them is to be accepted provisionally. Therefore, we must adopt the methodological rule that such statements should remain always open to critical discussion, even after its truth-value has been preliminarily determined.

Popper came to these conclusions between 1932 and 1933 and hastened to insert some in §11 of the text, but immediately realized that a drastic revision was needed. He consequently gave up the project of writing *Die beiden Grundprobleme der Erkenntnistheorie*, leaving only a few fragments of what was meant to be its second volume.[78] In any event, the remaining fragments are very important. A first group[79] is particularly worth considering, because it illustrates the transition from the study of scientific method as logical analysis to that understood as the study of methodological rules. It comprises two fragments, "Übergang zur Methodentheorie" ("Transition to the theory of Method") and "Das Problem der Methodenlehre" ("The Problem of the Doctrine of Method.")[80] The former, in particular, opens with what is nowadays usually referred to as "Duhem thesis," according to which scientific theories can never be refuted by experience, since it is always possible to modify them so as to "save" them from falsification. This objection may be replied to only by appealing to the methodological decision not to avoid refutations. A second group, comprising three short fragments,[81] aims at defending Popper's methodological proposal from the logical positivists' objections. Finally, a third group deals with a few problems raised by the appeal to methodology. Of the seven fragments it comprises, "Problemstellung" ("Definition of the Problem") offers a revision of the demarcation problem that switches from the problem of demarcating scientific statements to the problem of identifying empirical sciences,[82] whereas "Grundriss einer Theorie der empirisch-wissenschaftlichen Methoden (Theorie der Erfahrung)" ("Sketch f a Theory of the Empirical-Scientific Method [Theory of Experience])" proposes a solution of the problem of *ad hoc* hypotheses.[83]

The writing of *Die beiden Grundprobleme der Erkenntnistheorie* was therefore interrupted, and Popper devoted himself to a new project. In fact, *Logik der Forschung* was not a "radically shortened" (or however modified) version of *Die beiden Grundprobleme der Erkenntnistheorie*, as Popper said in his autobiography.[84] It was an entirely new work: only by introducing some methodological rules and only after abandoning the project of demarcating scientific statements from nonscientific ones on the basis of the proof of their truth or falsity could Popper conceive of a concrete and viable alternative to logical positivists' foundational theories.

POPPER'S PHILOSOPHICAL BREAKTHROUGH: *LOGIK DER FORSCHUNG*, 1934

In order to accomplish the new philosophy of science he intended to construct, Popper first restated the problems of induction and demarcation: they no longer refer to the way in which statements are proved or disproved, but rather to the methodological problems involved in dealing with scientific theories. However closely connected, the problem of induction does not coincide with the problem of demarcation any more.[85] Having given up the problem of determining the truth-value of statements in favour of methodological rules, Popper can now disregard the problem of the truth-value of statements that cannot be proved true or false, which in his previous attempt had led him to the unsatisfactory idea that universal statements are "fictions."

Furthermore, Popper's new view is closer to Einstein's. In the 1929 thesis Popper said that scientists choose that geometry as the best approximation to reality which, in combination with some physical theory, yields the simplest combination. He now combines this remark with his new theory of demarcation and with the considerations dealing with the choice to regard a theory as refuted: scientists provisionally adopt the theory with the highest degree of refutability; this will turn out to be the simplest.

The feverish and confused years from 1927 to 1934 mark the steps of young Popper's antifoundationalist intellectual revolution. He managed to construct a new philosophy of science "within" the Würzburg School of psychology by integrating the views of its exponents with his own. This filled the gap, in the development of the School, left by Külpe's failure to provide a consistently deductivist philosophy of science, although he himself felt the need for one. Popper also realized that in order to fill that gap he needed to learn the new logic and the new science. He learned what he could from Schlick, Carnap, Reichenbach, and other members of the Vienna Circle, thus giving his early attempts a decidedly positivist flavour. Confronted with their own failures, however, he abandoned their track and, thanks to the long discussions with Kraft, he decided to follow the footprints of Fries and Nelson. He learned from Poincaré and Einstein that science is not inductive, and from the latter that it is conjectural and aims at the truth.

This path was not easy at all, and it is amazing to see how Popper found his way in this maze of ideas: he gave up the inductivism he adopted in the 1927 thesis; he quickly learned how to employ the new logic without embracing an inductivist or a conventionalist stance; he discovered how to make use of some ideas of the Würzburg School while at the same time rejecting its methodological views (with the only exception some remarks by Selz); he rapidly understood the need to follow the model of Einstein, although he had not developed his views; and he realized that Einstein's views constituted a challenge for the traditional picture of the superiority of scientific knowledge. Vienna's philosophical establishment had not grasped the revolutionary philosophical import of the new physics. New answers were called for. Popper offered *Logik der Forschung*.

2 Science and Philosophy

As we have seen, the two fundamental problems of the theory of knowledge singled out by Popper in 1933 are the problem of induction—that is, "The question whether inductive inferences are justified, or under what conditions"[1]—and the problem of demarcation—that is, "the problem of finding a criterion which would enable us to distinguish between the empirical sciences on the one hand, and mathematics and logic as well as 'metaphysical' systems on the other."[2] Following Kant, Popper called the former "Hume's problem" and he thought it appropriate to refer to the latter as "Kant's problem."[3]

HUME'S PROBLEM

Induction—from the Latin *inductio*, Cicero's term to translate the Greek *epagogé*, with which Aristotle referred to the passage from the particular to the universal—was already validly criticized by Sextus Empiricus in his *Outlines of Pyrrhonism*: induction cannot justify the acceptance of universal statements as true. For this reason Hume distinguished the genesis of an expectation, or a hypothesis, and its validity. For, while discussing the idea of necessary connection (causality), the Scottish philosopher distinguished a logical and an empirical part of the problem. This is Popper's point of departure.

Hume's logical problem is the following: "Are we justified in reasoning from [repeated] instances of which we have experience to other instances [conclusions] of which we have no experience?"[4] Hume's answer is in the negative: we are not justified in doing so, however great the number of repetitions. Popper tells a well-known example: the repeated observation of many white swans does not allow us to conclude that all swans are white, since the logical content of the conclusion exceeds by far any particular statement referring to a limited number of observations, however great.[5]

On the other hand, the empirical problem (or psychological problem, as Popper labels it) can be stated as follows: "Why, nevertheless, do all reasonable people expect, and *believe*, that instances of which they have no experience will conform to those of which they have experience?"[6] The reason, Hume observed, is to be found in custom or habit: for we are conditioned by repetition and by the mechanism of the association of ideas—indeed, without it, we would not be able to survive.[7] Popper accepts Hume's negative opinion about the logical problem: induction is logically invalid and in no sense justified. But he rejects Hume's opinion about the psychological one, that is, that induction is a fact, and at any rate needed. His conclusion is clear-cut: "Induction simply does not exist, and the opposite view is a straightforward mistake."[8]

> The belief that we use induction is simply a mistake. It is a kind of optical illusion. What we do use is a method of trial and of the elimination of error; however misleadingly this method may look like induction, its logical structure, if we examine it closely, totally differs from that of induction. Moreover, it is a method which does not give rise to any of the difficulties connected with the problem of induction. Thus it is not because induction can manage without justification that I am opposed to the traditional problem; on the contrary, it would urgently need justification. But the need cannot be satisfied.[9]

The first step towards a positive solution of the problem of induction is the restatement of Hume's logical problem: "Can the claim that an explanatory universal theory is true be justified by 'empirical reasons?'"[10] Popper's answer is negative, just as Hume's was: "no number of true test statements would justify the claim that an explanatory universal theory is true,"[11] for "we must regard *all laws or theories as hypothetical or conjectural*; that is, as guesses."[12] Then Popper generalizes the problem: "Can the claim that an explanatory universal theory is true or that it is false be justified by 'empirical reasons?'"[13] To this second question, Popper answers affirmatively: "Yes, *the assumption of the truth of test statements sometimes allows us to justify the claim that an explanatory universal theory is false.*"[14] Moreover, the problem of the choice among different explanatory theories competing for the solution of a given problem leads Popper to a further restatement of the logical problem of induction: "Can a *preference*, with respect to truth or falsity, for some competing universal theories over others ever be justified by such 'empirical reasons?'"[15] In the light of the previous affirmative answer, this latter question can be replied affirmatively as well: "Yes; sometimes it can, if we are lucky. For it may happen that our test statements may refute some—but not all—of the competing theories; and since we are searching for a true theory, we shall prefer those whose falsity has not been established."[16]

The work of the scientists, Popper says, consists in the elaboration of theories and in putting them to the test. The initial phase, in which theories are conceived, does not require a logical analysis. Popper is very clear about this—the process of devising a new idea and the methods and results of its test are distinct:

> The question how it happens that a new idea occurs to a man—whether it is a musical theme, a dramatic conflict, or a scientific theory—may be of great interest to empirical psychology; but it is irrelevant to the

logical analysis of scientific knowledge. This latter is concerned not with *questions of fact* (Kant's *quid facti?*), but only with questions of *justification or validity* (Kant's *quid juris?*).[17]

The method of critically testing theories, and then selecting them according to the results of tests, proceeds as follows:

> From a new idea, put up tentatively, and not yet justified in any way—an anticipation, a hypothesis, a theoretical system, or what you will—conclusions are drawn by means of logical deduction. These conclusions are then compared with one another and with other relevant statements, so as to find what logical relations (such as equivalence, derivability, compatibility, or incompatibility) exist between them.[18]

Testing a theory against experience takes place through the deduction of predictions: "if the singular conclusions turn out to be acceptable, or *verified*, then the theory has, for the time being, passed the test: we have found no reason to discard it. But [. . .] if the conclusions have been *falsified*, then their falsification also falsifies the theory from which they were logically deduced."[19] This is an important point:

> a positive decision can only temporarily support the theory, for subsequent negative decisions may always overthrow it. So long as a theory withstands detailed and severe tests and is not superseded by another theory in the course of scientific progress, we may say that is has "proved its mettle" or that it is "*corroborated*" by past experience.[20]

KANT'S PROBLEM

In the years immediately following World War I, Austria was pervaded by new and revolutionary ideas. Popper was particularly attracted by Einstein's theory of relativity (which received the first crucial empirical support in 1919), Marx's theory of history, Freud's psychoanalysis, and Adler's "individual psychology." As years went by, however, Popper felt increasingly dissatisfied with the latter three:[21] "what worried me was neither the problem of truth, at that stage at least, nor the problem of exactness or measurability. It was rather that I felt that these other three theories, though posing as sciences, had in fact more in common with primitive myths than with science; that they resembled astrology rather than astronomy."[22] In particular, Popper highlights the alleged explanatory power of Marx's theory, Freudian psychoanalysis, and Adler's psychology:

> These theories appeared to be able to explain practically everything that happened within the fields to which they referred. The study of any of them seemed to have the effect of an intellectual conversion or

revelation, opening your eyes to a new truth hidden from those not yet initiated. Once your eyes were thus opened you saw confirming instances everywhere: the word was full of verifications of the theory. Whatever happened always confirmed it.[23]

Things were different in the case of Einstein's theory, which made precise, risky predictions, that were incompatible with some possible experimental results. Popper's conclusion is that: "It is easy to obtain confirmations, or verifications, for nearly every theory—if we look for confirmations [. . .] Irrefutability is not a virtue of a theory (as people often think) but a vice."[24]

What characterizes science, that is, is the possibility of being refuted: as opposed to psychoanalysis or Marxist historicism, science is susceptible of being contradicted by experience. This is the very point inductive logic cannot grasp: "*it does not provide a suitable distinguishing mark* of the empirical, non-metaphysical, character of a theoretical system; or, in other words, [. . .] *it does not provide a suitable 'criterion of demarcation'.*"[25] Therefore, Popper's solution of the problem of demarcation relies on the logical consideration that universal statements are in no way verifiable.[26] In the example mentioned earlier: no finite series of observations of white swans can ever verify the truth of the general conclusion that all swans are white. Whereas no universal statements can be verified by experience, given its necessary finiteness, any universal statements can be falsified by experience, provided we assume the truth of at least one basic statement, that is, if there exists at least one observation statement that conflicts with the universal proposition. The observation of even one single black swan clashes with, and therefore falsifies, the universal statement "all swans are white." As a consequence, there is "an asymmetry between verifiability and falsifiability; an asymmetry which results from the logical form of universal statements. For these are never derivable from singular statements, but can be contradicted by singular statements."[27]

Faced with the impossibility of empirically verifying theories, Popper advances a *criterion of demarcation* in order to distinguish empirical science from "metaphysics." On the basis of this criterion, "a system [will be admitted] as empirical or scientific only if it is capable of being *tested* by experience."[28] Scientific character is thus attributed on the basis of the *falsifiability*, as opposed to the verifiability, of a theoretical system.[29]

In other words: I shall not require of a scientific system that it shall be capable of being singled out, once and for all, in a positive sense; but I shall require that its logical form shall be such that it can be singled out, by means of empirical tests, in a negative sense: *it must be possible for an empirical scientific system to be refuted by experience.*[30]

All statements that are neither analytic (and thus may belong to pure sciences), nor contradictory, nor empirically falsifiable, belong to metaphysics. By "metaphysics," then, Popper simply means a category comprising everything that does not belong to science according to Popper's criterion of demarcation. This does not mean that it is meaningless or bears no value for science proper.[31]

In order to be empirical, a theory must be falsifiable: it must be able to divide all possible basic statements—the elements, that is, through which theory connects with reality and hence can be tested[32]—unambiguously into two nonempty classes. The first is the class of the basic statements with which the theory clashes, or is inconsistent (those basic statements the theory excludes, or rules out): this is the class of the theory's *potential falsifiers*. The second class comprises those basic statements which the theory does not contradict (or permits). The empirical character of a theory can therefore be expressed as follows: "a theory is falsifiable [*i.e.*, empirical] if the class of its potential falsifiers is not empty."[33]

In order to better specify his proposed criterion of demarcation, Popper devotes the whole of Chapter 6 of *The Logic of Scientific Discovery* to "Degrees of Testability": for, he says, falsifiability is "a matter of degree."[34] If the class of the potential falsifiers of a theory is larger than that of another one, the former is more likely to be refuted by experience than the latter. It will have a higher degree of testability (or falsifiability)—it will tell us more about the world, since it excludes a greater number of basic statements: "the amount of empirical information conveyed by a theory, or its *empirical content*, increases with its degree of falsifiability."[35]

Popper's proposal[36] contrasts with the preceding criteria of demarcation, based upon meaning. The idea that metaphysics is devoid of meaning could be traced back to Thomas Hobbes, George Berkeley, and David Hume, and more recently to Charles Sanders Peirce (to whom Quine attributes the verification theory of meaning[37]). Popper's target, however, was Ludwig Wittgenstein and the members of the Vienna Circle. Whereas Hume considered the meaning of words and Peirce that of concepts and other constructs, Wittgenstein was concerned with the meaning of propositions:[38]

Philosophy aims at the logical clarification of thoughts. [. . .]
A philosophical work consists essentially of elucidations.
Philosophy does not result in "philosophical propositions," but rather in the clarification of propositions.
Without philosophy thoughts are, as it were, cloudy and indistinct: its task is to make them clear and to give them sharp boundaries.[39]

Philosophy settles controversies about the limits of natural science.[40]

It must set limits to what can be thought; and, in doing so, to what cannot be thought.
It must set limits to what cannot be thought by working outwards through what can be thought.[41]

It will signify what cannot be said, by presenting clearly what can be said.[42]

The criterion of demarcation advanced by positivists—the original formulation is due to Moritz Schlick and Friedrich Waismann, in the footsteps of Wittgenstein[43]—was first proposed as a criterion of meaning: we can regard as scientific all and only those statements that can be verified by observation statements; these propositions coincide, in their turn, with the set of meaningful statements.[44] The positivists' aim, however, was not merely an effective criterion of demarcation, but rather—in Carnap's own words—the overcoming, or elimination of metaphysics, its ultimate exclusion from science, properly understood.[45] However, in Popper's eyes the criterion advanced by Wittgenstein and the positivists was "totally inadequate."[46] For, it "is too narrow (*and* too wide): it excludes from science practically everything that is, in fact, characteristic of it (while failing in effect to exclude astrology)."[47] The criterion of verification ends up excluding, as nonverifiable, all universal statements—and therefore all natural laws, that science aims at discovering and that are characteristically universal; at the same time, however, it includes other disciplines, such as astrology (or draws from a lottery), that we would want to exclude from science: for such predictions may get verified. On the one hand, then, positivists "in their anxiety to annihilate metaphysics, annihilate natural science along with it";[48] on the other, "instead of eradicating metaphysics from the empirical sciences, positivism leads to the invasion of metaphysics into the scientific realm."[49]

THE ROLE OF METAPHYSICS

One of the most important aspects of Popper's criticism of the members of the Vienna Circle concerns the role of metaphysics: not only is it meaningful, but it can also play a positive role for science. Metaphysics, so to say, is the fluid environment within which proper science is nourished, the fertile soil from which scientific ideas spring.[50] For, as time goes by, some metaphysical ideas may well turn into scientific hypotheses:

> To obtain a picture or model of this quasi-inductive evolution of science, the various ideas and hypotheses might be visualized as particles suspended in a fluid. Testable science is the precipitation of these particles at the bottom of the vessel: they settle down in layers (of universality). The thickness of the deposit grows with the number of these layers, every new layer corresponding to a theory more universal than those beneath it. As the result of this process ideas previously floating in higher metaphysical regions may sometimes be reached by the

growth of science, and thus make contact with it, and settle. Examples of such ideas are atomism; the idea of a single physical "principle" or ultimate element (from which the others derive); the theory of terrestrial motion (opposed by Bacon as fictitious); the age-old corpuscular theory of light; the fluid-theory of electricity (revived as the electron-gas hypothesis of metallic conduction).[51] All these metaphysical concepts and ideas may have helped, even in their early forms, to bring order into man's picture of the world, and in some cases they may even have led to successful predictions.[52]

Among the various examples offered by Popper, atomism is perhaps the most effective: the idea that matter is constituted by indivisible particles (atoms) was advanced by Leucippus and Democritus in the fifth century BC, later to spread and become popular thanks especially to Epicurus in Greece and Lucretius in Rome. It gained new vigour in the sixteenth and seventeenth centuries, until John Dalton, at the beginning of the nineteenth century, reworked it in order to solve a few problems he was facing in chemistry. Later on, James Clerk Maxwell introduced atomism into mathematical physics in connection with the kinetic theory of gases. At the time it was introduced, atomism was clearly not testable, and hence not a scientific hypothesis—a sort of grand "theory of everything." However, at the beginning of the twentieth century, it turned to be a scientific hypothesis in every respect. Atoms as understood my physicists nowadays are certainly very different from those Leucippus and Democritus had conceived of some twenty-five centuries ago, but modern physics could have hardly spoken of atoms the way it does were it not for the preceding metaphysical atomistic tradition:

> In fact, the example of atomism established the inadequacy of the doctrine that metaphysics is mere meaningless gibberish. And it establishes the inadequacy of the policy of making little surreptitious changes here and there to the doctrine of meaninglessness, in the vain hope of rescuing it. [. . .] atomism is an excellent example of a non-testable metaphysical theory whose influence upon science has exceeded that of many testable scientific theories.[53]

In the first volume of the *Postscript* Popper develops the view of metaphysics outlined in *The Logic of Scientific Discovery*, introducing the idea of a "metaphysical research programme."[54] Before becoming testable, general views such as atomism played the heuristic role of a research programme for science, pointing to the directions along which research might develop, or the form satisfactory explanations might assume.

However important, metaphysical programmes are not to be confused with testable theories. Yet, it is possible, in Popper's view, to assess an irrefutable theory rationally: "a critical discussion even of irrefutable theories may well be possible."[55] For, although nonempirical and nontestable, we can nevertheless ask questions such as: "Does it solve the problem? Does it solve it better than other theories? Has it perhaps merely shifted the problem? Is the solution simple? Is it fruitful? Does it perhaps contradict other philosophical theories needed for solving other problems?"[56] Being empirically nontestable, but critically testable, metaphysical theories entertain a close connection with scientific thought: for, from the historical point of view, metaphysics provides the fertile soil from which scientific theories spring up; and, from the heuristic point of view, it provides scientists with fundamental and important regulative ideas as long as, by conveying ways of conceiving of the world, it suggests new methods for exploring it.

Metaphysical research programmes is one of Popper's ideas best developed by his pupils and critics: John Watkins speaks of an "influential metaphysics" influencing science from the outside;[57] Joseph Agassi suggests viewing "some metaphysics as the possible foundation of future science; [. . .] as often conflicting with existing scientific theories and as incentives to alterations which would remove the conflict."[58] Pushing further than Thomas Kuhn, who speaks of "metaphysical paradigms" or "the metaphysical parts of paradigms,"[59] Imre Lakatos locates metaphysics in the hard core of his scientific research programmes, that is, at the very heart of the scientific enterprise.[60]

METHODOLOGICAL RULES AND THE IMPORTANCE OF DECISIONS

A number of objections might be raised against testability as the criterion of demarcation, and it is Popper himself who discusses some of them after introducing it. Let us consider just two of them here. The first: "it may well seem somewhat wrong-headed to suggest that science, which is supposed to give us positive information, should be characterized as satisfying a negative requirement such as refutability."[61] This objection has little weight, though, "since the amount of positive information about the world which is conveyed by a scientific statement is the greater the more likely it is to clash, because of its logical character, with possible singular statements."[62] Our theories, in other words, have the greater content the more they prevent possible states of affairs: "the more they prohibit the more they say."[63]

The second, more substantial objection comes from conventionalism. In the eyes of a conventionalist, the seeming simplicity of the world, as revealed in the laws of physics, would be utterly incomprehensible were we bound to believe, with the realists, that the laws of nature reveal an inner, structural simplicity of the world underlying the outer appearances of phenomena. Such simplicity, conventionalists argue, is due to the fact that so-called natural laws are, in fact, our own creations. If Kant's idealism sought to explain this simplicity by saying that it is our own intellect which imposes its laws upon nature, conventionalists argue that the laws of nature are arbitrary conventions, produced by our own free decisions. As a consequence, natural science would not provide a picture of the world as it is, but, rather, a mere logical reconstruction of it: "It is not the properties of the world which determine this construction; on the contrary it is this construction which determines the properties of an artificial world: a world of concepts implicitly defined by the natural laws which we have chosen."[64] In the eyes of the conventionalists, then, the laws of nature are not falsifiable by observations, since it is the laws themselves that determine the observations. Partly resuming the analysis developed in his 1929 *Habilitation* thesis about the history of non-Euclidean geometries and the applicability of axiomatic-deductive systems to reality, Popper acknowledges how conventionalism—as opposed to inductivism—has contributed to clarifying the relationship between theory and experiment.[65] And yet, in spite all of that, he finds conventionalism "quite unacceptable":[66] for, whilst Popper does not demand any final certainty for science, conventionalists consistently seek an ultimate foundation for knowledge.[67]

While recognizing the asymmetry between verification and falsification, conventionalists might object to Popper that just as it is not possible to obtain a conclusive verification of theoretical systems in natural sciences, so it is impossible to conclusively falsify them.[68] For it is always possible "to save the phenomena" by introducing *ad hoc* hypotheses, modifying ostensive definitions, doubting the reliability of the experimenter, or rejecting any falsifying instances. Form a purely logical point of view, that is, Popper's proposed criterion of demarcation does not lead to an unambiguous classification of theoretical systems into falsifiable and non-falsifiable ones—and hence fails to achieve its aim. Popper acknowledges the validity of this objection, but responds by calling attention to the need to *integrate logic and methodology*.[69] Epistemology "should be identified with the theory of scientific method."[70] Such a theory, "in so far as it goes beyond the purely logical analysis of the relations between scientific statements, is concerned with *the choice of methods*—with decisions about the way in which scientific statements are to be dealt with."[71]

> The question whether a given *system* should as such be regarded as a conventionalist or an empirical one is [. . .] misconceived. *Only with reference to the methods applied* to a theoretical system is it at all possible to ask whether we are dealing with a conventionalist or an empirical theory. The only way to avoid conventionalism is by taking a *decision*: the decision not to apply its methods. We decide that if our system is threatened we will never save it by any kind of *conventionalist stratagem*.[72]

Of course, these decisions depend on the goal we want to achieve when doing research. In connection with his proposal for a criterion of demarcation, Popper proposes "to adopt such rules as will ensure the testability of scientific statements; which is to say, their falsifiability."[73]

[E]*mpirical method* shall be characterized as a method that excludes precisely those ways of evading falsification which [. . .] are logically possible. According to my proposal, what characterizes the empirical method is its manner of exposing to falsification, in every conceivable way, the system to be tested. Its aim is not to save the lives of untenable systems but, on the contrary, to select the one which is by comparison the fittest, by exposing them all to the fiercest struggle for survival.[74]

On the basis of the failure of his attempt in the first volume of *Die beiden Grundprobleme der Erkenntnistheorie* (without §11)—in which he tried to develop a deductivist theory from the assumption that the aim of science is proof—Popper resolutely asserts that "If you insist of strict proof (or strict disproof) in the empirical sciences, you will never benefit from experience, and never learn from it how wrong you are."[75]

Popper regards methodological rules as conventions: "They might be described as the rules of the game of empirical science. They differ from the rules of pure logic rather as do the rules of chess."[76] Just as the game of chess might be defined by its rules, so empirical science is defined by the rules of its method:

> The game of science is, in principle, without end. He who decides one day that scientific statements do not call for any further test, and that they can be regarded as finally verified, retires from the game.[77]

The first and most important rule sets a general code of conduct: "It is the rule which says that the other rules of scientific procedure must be designed in such a way that they do not protect any statement in science against falsification."[78] Its aim is to guarantee that other rules improve the applicability of the criterion of demarcation.

Another rule states that "only such statements may be introduced in science as are inter-subjectively testable."[79] This is another important point: intersubjectivity, that is, independent control by different people, replaces objectivity, which remains unattainable.[80] Theories can never be justified, but can be subjected to independent tests: "the *objectivity* of scientific statements lies in the fact that they can be *inter-subjectively tested*."[81]

As to auxiliary hypotheses, Popper lays down the rule that "only those are acceptable whose introduction does not diminish the degree of falsifiability or testability of the system in question, but, on the contrary, increases it."[82] In other words, the introduction of an auxiliary hypothesis strengthens the theoretical system only if it increases the chances to put it to the test and possibly refute it; otherwise, it can only weaken it. In the latter case the hypothesis in question has been introduced *ad hoc*, with the only aim of avoiding possible falsifications. "The introduction of an auxiliary hypothesis," Popper explains—thus implicitly countering the so-called "Duhem thesis"[83]—"should always be regarded as an attempt to construct a new system; and this new system should then always be judged on the issue of whether it would, if adopted, constitute a real advance in our knowledge of the world."[84]

THE PROBLEM OF THE "EMPIRICAL BASIS"

If we are to apply the falsifiability criterion, we need singular statements that may serve as premises of falsifying inferences. It is on this very issue—crucial for Popper's falsificationism—that we see the culmination of the revolution whose steps we quickly followed in the previous chapter. In *Neue oder anthropologische Kritik der Vernunft* (1828–1831) Fries posed the problem in clear-cut terms:[85] if we do not wish to accept scientific statements dogmatically, we should be able to justify them; and if we require justification by way of logical arguments, then we need to accept that statements can only be justified by other statements. The requirement that all statements are justified necessarily leads to infinite regress. Therefore, in order to avoid both the danger of dogmatism and that of infinite regress, we can only opt for psychologism: statements can be justified not only by other statements, but also by perceptual experience.

Fries opted for psychologism, as did most of the philosophers of science who accepted the assumption that scientific knowledge, as such, must be founded. Among these were Otto Neurath and Rudolf Carnap, although instead of "experience" and "perceptions" they preferred to speak of "statements" that represent experiences: the so-called protocol statements, or protocols.[86] Protocols are statements needing no justification, but rather serving as foundation for all remaining statements of science. They "translate" facts into words, directly describing given experience or phenomena, that is, the simplest states of which knowledge can be had.[87]

Unlike Fries and the logical positivists, Popper cut the Gordian knot posed by the trilemma and did not opt for any of its alternatives.[88] All tests to which a theory is put—those with a positive result as well as those with a negative one—must stop at statements we decide to accept. We are not obliged to accept them, or to give up testing: any basic statements can in turn be put to the test, by appealing to other basic statements as touchstones. The process of testing has no natural end: we stop when we are satisfied, at least for the time being; however, we can resume the testing process at any time.

The basic statements at which we stop have admittedly the character of dogmas, so to say, but not in the sense that we give up testing—it is an innocuous form of dogmatism, that is: we could restart the testing process whenever we think it appropriate. Thus understood, the chain of deductions and tests is virtually infinite and would lead to infinite regress—but such an infinite regress is innocuous too, since in Popper's proposal there is no question of trying to prove any statement by means of it. Finally, the decision to accept a basic statement is certainly connected with our perceptual experiences—but, again, our aim is not justifying basic statements by means of such experiences.[89] Dogmatism, infinite regress, and psychologism all play a role in scientific research, but are rendered innocuous by the hypothetical and falsifiable character of science.

Our preference for one theory over another is not due to any experimental justification:

> We choose the theory which best holds its own in competition with other theories; the one which, by natural selection, proves itself the fittest to survive. This will be the one which not only has hitherto stood up to the severest tests, but the one which is also testable in the most rigorous way. A theory is a tool which we test by applying it, and which we judge as to its fitness by the results of its applications.[90]

The test of a theory depends on basic statements, whose acceptance or rejection is up to us to decide. Such a decision, on our part, may be partly due to considerations of utility, as conventionalists say. There is a "vast difference" between Popper's views and the conventionalists, for Popper holds that "what characterizes the empirical method is just this: that the convention or decision does not immediately determine our acceptance of *universal* statements but that, on the contrary, it enters into our acceptance of the *singular* statements—that is, the basic statements."[91] The conventionalists' decision to accept universal statements is governed by a principle of utility, or simplicity;[92] for Popper, by contrast, "the first thing to be taken into account should be the severity of tests. [. . .] what ultimately decides the fate of a theory is the result of a test, *i.e.* an argument about basic statements."[93]

> The empirical basis of objective science has thus nothing "absolute" about it. Science does not rest upon solid bedrock. The bold structure of its theories rises, as it were, above a swamp. It is like a building erected on piles. The piles are driven down from above into the swamp, but not down to any natural or "given" base; and if we stop driving the piles deeper, it is not because we have reached firm ground. We simply stop when we are satisfied that the piles are firm enough to carry the structure, at least for the time being.[94]

CORROBORATION

Once the hope of proving the truth or falsity of theories has been given up, many philosophers have made do with their probability. Besides the values "true" and "false," in the eyes of many inductive logic can assign statements degrees of probability between 0 (impossibility) and 1 (certainty) as well.[95] Thus stated, however, the whole question is misconceived, according to Popper: "Instead of discussing the 'probability' of a hypothesis we should try to assess what tests, what trials, it has withstood; that is, we should try to assess how far it has been able to prove its fitness to survive by standing up to tests. In brief, we should try to assess how far it has been 'corroborated.'"[96]

A theory is said to be *corroborated* as long as it withstands its tests.[97] The assessment of corroboration establishes relations such as the compatibility or incompatibility of the theory with one of its potential falsifiers. If incompatibility amounts to falsification, and therefore the decision to accept it implies the decision to regard the theory as falsified,

> compatibility alone must not make us attribute to the theory a positive degree of corroboration: the mere fact that a theory has not yet been falsified can obviously not be regarded as sufficient. For nothing is easier than to construct any number of theoretical systems which are compatible with any given system of accepted basic statements.[98]

For, what determines the degree of corroboration is not so much the number of corroborating instances, "as the severity of the various tests to which the hypothesis in question can be, and has been, subjected."[99]

As a consequence, "the corroborability of a theory—and also the degree of corroboration of a theory which has in fact passed severe tests, stand both, as it were, in inverse ratio to its logical probability; for they both increase with its degree of testability and simplicity."[100] This is the key point of Popper's criticism of the probabilistic theory of induction. We want simple hypotheses, that is, hypotheses with a high content, or a high degree of testability.[101] These are also highly corroborable hypotheses, for the degree of corroboration of a hypothesis depends (mainly) upon the severity of tests, and thus upon its testability. But high testability—as opposed as to Carnap's and Reichenbach's contentions—equals high (absolute) logical improbability, or low (absolute) logical probability. Let us assume that $p(a)<p(b)$, where $p(a)$ and $p(b)$ are the (absolute) probabilities of two hypotheses, a and b; and let $p(a|e)$ and $p(b|e)$ denote the *a posteriori* probabilities of the two hypotheses a and b, that is, the (relative, conditional) logical probability of a and b given the evidence e. Then, for any given e, we will have $p(a|e)<p(b|e)$. "Thus the *better testable and better corroborable hypothesis can never obtain a higher probability, on the given evidence, than the less testable one. But this entails that degree of corroboration cannot be the same as probability.*"[102] The view implied by probability logic is the precise opposite of Popper's own: for its upholders let the probability of a hypothesis increase in direct proportion to—as opposed to in inverse proportion to, as Popper argues—proportion to its logical probability.[103] Therefore, high probability cannot be the aim of science, since improbability is associated with content: "*the more a statement asserts, the less probable it is.*"[104]

The *ratio negativa* of the criterion of falsifiability finds its complement in the idea of corroboration, in the positive support provided by experience to theories. Indeed, corroboration grows out of the very problems of induction and demarcation.[105] The idea of the degrees of corroboration, Popper explains, "was to sum up, in a short formula, a *report* of the manner in which a theory has passed—or not passed—its tests, including an evaluation of the severity of the tests [. . .]. By passing such tests, a theory may 'prove its mettle'—its 'fitness to survive.'"[106] This is, however, only a critical report on the quality of past performances, that cannot be used to predict future performances:

> just as in the case of an organism, "fitness," unfortunately, only means actual survival, and past performance in no way ensures future success. [. . .] one could only speak of the degree of corroboration of a theory *at a certain stage of its critical discussion*. In some cases it provided a very good guide if one wished to assess *the relative merits of two or more competing theories in the light of past discussions*. When faced with the *need to act*, on one theory or another, the rational choice was to act on that theory—if there was one—which so far has stood up to criticism better than its competitors had.[107]

The empirical support a theory gains from experience is only a measure of its having stood up to severe tests—but no degree of corroboration of a hypothesis can secure that it will not be refuted in the next test; no degree of corroboration of a hypothesis makes it even slightly less probable that it will be refuted in the next test. Corroboration implies neither verification nor any increase of probability. Degree of corroboration serves as a rational guide to practice: since it is not possible to justify our belief in the truth of a theory, we can sometimes justify our preference for one theory over another. At first sight, corroboration seems quite commonsensical and unproblematic—a mere addition to the picture of science Popper depicts in *The Logic of Scientific Discovery*. The issue is quite important, though, and has raised some discussion, which is worth summarizing here.

In *The Logic of Scientific Discovery* Popper proposes that a theory is scientific if and only if it can be overthrown with the help of experience, and that we gain theoretical knowledge from experience when and only when revolutions occur. There is nothing explicit about corroboration, namely about the failure to refute hypotheses.[108] Popper's solution of the problem of induction has nothing to do with his theory of corroboration, which came later, but hinges on his theory of gradual approximation to the truth by repeatedly making explanatory hypotheses and refuting them experimentally. This solution offers a model of the progress of science and of the growth of knowledge by a never-ending process of conjectures and refutations—of bold scientific hypotheses which we must do our best to criticize and refute. As we have seen, in *The Logic of Scientific Discovery* Popper declares learning from experience to be not by positive evidence but by negative evidence. For, if we search for positive evidence, the evidence will not be scientific; the only way to find positive evidence recognizable by science is by looking for negative evidence. If we look for negative evidence, says Popper, we may find it, or else we will find positive evidence; but if we look for positive evidence we will find only positive evidence, which is of no scientific value.

Popper views science as a special case of Socratic dialogue, with experiments and observations offering new arguments, or new empirical criticisms. And when an attempt at empirical criticism misfires the result is positive evidence. In *The Logic of Scientific Discovery* corroboration is only the measure of a test gone wrong. Whatever role positive evidence may play, it cannot play any role *qua* positive evidence. It may stimulate the invention of a conjecture and it may be not only a positive support of one conjecture, but also a refutation of another. This, but no more: the corroboration of a theory is enlightening, though less enlightening than its refutation—but it is not really necessary.

It is only with "Truth, Rationality, and the Growth of Scientific Knowledge"—a paper first published as Chapter 10 of *Conjectures and Refutations*, in 1963—that Popper claims that corroboration is as important and not merely as enlightening as discovery and increasing of explanatory power of a hypothesis.[109] Corroboration, he now says, is essential to science, since it provides an encouragement to our researches: a good theory, he argues, should be not only capable of being refuted, but it should not be refuted too soon. We want positive evidence before we get negative evidence, so as to be assured that knowledge grows.

In so doing, Popper changes his view of science as conjectures and refutations into a view of science as conjectures, corroborations, and refutations. For in his 1963 paper—which, as he declares, "contains [. . .] some essential developments of the ideas of my *Logic of Scientific Discovery*"[110]—he introduces a "third requirement" that a new scientific theory must satisfy if it is to be regarded as an advancement. The first two requirements were: (a) "The new theory should proceed from some *simple, new, and powerful, unifying idea* about some connection or relation [. . .] between hitherto unconnected things [. . .] or facts [. . .] or new 'theoretical entities'"; and (b) "the new theory should be *independently testable*."[111] The third is: "the theory should pass some new, and severe, tests."[112] This latter requirement may be divided into two parts: "first we require of a good theory that it should be successful in some of its new predictions; secondly we require that it is not refuted too soon—that is, before it has been strikingly successful."[113]

As some critics have remarked, here Popper seems to be changing the rules during the game, at the risk of falling back into the very justificationism he freed himself from in *The Logic of Scientific Discovery*.[114] By contrast, as Popper himself notices in his 1974 rejoinder to Alfred J. Ayer, "there might be a 'whiff' of inductivism here. It enters with the vague realist assumption that reality, though unknown, is in some respects similar to what science tells us or, in other words, with the assumption that science can progress towards greater verisimilitude."[115] And again, in a rejoinder to a paper by Agassi:[116] "I admit that there may be a whiff of verificationism here; but this seems to me a case where we have to put up with it, if we do not want a whiff of some form of instrumentalism that takes theories to be mere instruments of exploration."[117] These passages have been differently understood. Of the first, William Newton-Smith says that "it is just false to say that there is a whiff of inductivism here—there is a full-blown storm."[118] Of the second, David Miller says that "Popper's famous third requirement has to be seen as embodying not a 'whiff of verificationism' [. . .], but what might be called a whiff of verisimilitudinism."[119] By contrast, Joseph Agassi states that "either Popper assigns no value to positive evidence *qua* positive evidence, or he is in the same boat as the inductive philosophers who cannot bring positive evidence to support their theories of positive evidence."[120]

In fact, if our aim is to learn from experience all this is redundant. We learn from experience by repeatedly positing explanatory hypotheses and refuting them experimentally, thus approximating the truth by stages. We learn, that is, by criticizing our errors. From this point of view corroborations are entirely irrelevant. On the contrary, testing matters, and it has an undeniable epistemological significance. "Testing is important because it is only by subjecting our theories to tests that we have any opportunity of eliminating those that are false; and the more severe the test, the more generous the opportunity."[121] Corroborations are needed if science is to exist, for they illustrate the high explanatory power of the corrobarated hypothesis, but they make no contribution to the progress of science or to the growth of knowledge.

VERISIMILITUDE

The advancement of science is not due to the accumulation of new perceptual experiences, nor is it due to the fact that we are making ever better use of our senses: "Out of uninterpreted sense-experiences science cannot be distilled, no matter how industriously we gather and sort them. Bold ideas, unjustified anticipations, and speculative thought, are our only means for interpreting nature: our only organon, our only instrument, for grasping her."[122] We must make use of these means, throw ourselves into research and run the risk of committing mistakes: "Those among us who are unwilling to expose their ideas to the hazard of refutation do not take part in the scientific game."[123] Far from being the *tabula rasa* of the empiricists, our mind is a searchlight that casts its light upon the world. It has an active role and a highly selective task: "a scientific description will depend, largely, upon our point of view, our interests, which are as a rule connected with the theory or hypothesis we wish to test; although it will also depend upon the facts described."[124]

To the empiricist view, developed at least since Francis Bacon, of the mind as an empty bucket to be filled with the contents of experience, Popper opposes the view, of Kantian origin, of the mind as a searchlight that casts its light (hypotheses, theories, expectations) in the attempt to grasp reality more and more distinctly.[125] This is the sense of the words from

Novalis Popper chose as the motto for *The Logic of Scientific Discovery*: "Theories [from the 1992 reprint onwards, 'Hypotheses,' as in the German original] are nets: only he who casts will catch."[126] Theories are human inventions, not mere instruments. They are rational nets, designed by man to capture the world: "Theories are nets cast to catch what we call 'the world': to rationalize, to explain, and to master it. We endeavour to make the mesh ever finer and finer."[127] We aim at objective truth, in Tarski's sense, that is, at correspondence with the facts; its role is that of a regulative idea, as Kant or Peirce would have said.[128]

With Popper the century-old ideal of *episteme*, of absolutely certain and demonstrable knowledge, collapses: science is not a system of well-established statements, nor is it a system that steadily advances towards a state of finality. Our knowledge "can never claim to have attained truth, or even a substitute for it, such as probability."[129]

> The old scientific ideal of *episteme*—of absolutely certain, demonstrable knowledge—has proved to be an idol. The demand for scientific objectivity makes it inevitable that every scientific statement must remain *tentative for ever.*[130]

This is not a sceptical outcome—at least if we take *sképsis* in its original meaning, as we find it in Sextus Empiricus[131]—nor an irrational one: for if certainty is unattainable, the new fundamental problem of the theory of knowledge, that is, how it is possible for science to advance, and how to achieve progress, has a positive solution.[132] The search for truth remains at the heart of the scientific enterprise.

History of science seems to be a succession of falsified theories, and even the best theories available today—even the best corroborated ones—are bound to be replaced by others in a more or less distant future. Hence, we must confront the problem of how to assess the advancement of science. In order to solve this problem, Popper introduces the notion of verisimilitude (or truthlikeness):[133] the growth of science is measured by its ever better approximation to the truth.[134]

If we denote a measure of the "truth content" of the theory a, that is, the class of true logical consequences of a, with $Ct_V(a)$; and a measure of the "falsity content" of a, that is, the class of its false logical consequences, with $Ct_F(a)$, Popper defines verisimilitude as the difference of these two measures of content:

$$Vs(a) = Ct_V(a) - Ct_F(a)$$

Given two theories, t_1 and t_2, with comparable truth and falsity contents, we can say that t_2 is closer to the truth than t_1, that is, it better corresponds to facts, if: the truth content of t_2, but not its falsity content, exceeds that of t_1; or the falsity content of t_2, but not its truth content, exceeds that of t_2.[135] This definition, however, turned out to be wrong, for it fails to establish, between two false theories, which has a greater degree of verisimilitude, or is closer to the truth. As a result of the criticism raised independently by David Miller, Pavel Tichý, and John Harris in the 1970s,[136] Popper modified his original definition by stating, simply, that theory t_2 is closer to the truth than t_1 if and only if the (relative) truth content of t_2 exceeds the truth content of t_1, and some of the false consequences of t_1 can no longer be derived from t_2. Mistakes and logical difficulties of a formal definition of verisimilitude notwithstanding, Popper retained the idea of approximation to the truth, deeming it important—albeit not essential—to his own theory.[137]

Verisimilitude, as well as the truth (or falsity) content cannot be determined numerically, save the two cases of impossibility and certainty. Popper's aim

> is to achieve (on a lower level of precision) for verisimilitude something similar to what Tarski achieved for truth: the rehabilitation of a commonsense notion which has become suspect, but which is in my opinion much needed for any critical commonsense realism and for any critical theory of science. [. . .] In other words, my aim is the rehabilitation of a commonsense idea which I need for describing the aims of science, and which, I assert, underlies as a regulative principle (even if merely unconsciously or intuitively) the rationality of all critical scientific discussions.[138]

Even before a theory is put to the test, Popper argues, it is possible to know that, once it has passed a number of checks, it will be better than another theory. In order to do that, we need a criterion of relative potential satisfactoriness, or potential progressiveness, of a theory: "It characterizes as preferable the theory which tells us more; that is to say, the theory which contains the greater amount of empirical information or *content*; which is logically stronger; which has the greater explanatory and predictive power; and which can therefore be *more severely tested* by comparing predicted facts with observations. In short, we prefer an interesting, daring, and highly informative theory to a trivial one."[139] Our aim is truth, but not any truth. We are not interested in trivial truths—we want interesting truth, and therefore prefer informative theories to trivial ones.[140]

Writing *Ct(a)* and *Ct(b)* the contents of two statements a and b, respectively (for example: "it will rain on Friday" and "it will be sunny on Saturday"), it is clear that the content—or the quantity of information—of the conjunction ab of the two statements ("it will rain on Friday and it will be sunny on Saturday") will be greater than that of a or b alone:

$$Ct(a) \leq Ct(ab) \geq Ct(b)$$

According to the laws of the calculus of probability, however, the probability of the individual statements is higher than that of their conjunction:

$$p(a) \geq p(ab) \leq p(b)$$

The two relations have inverted inequality signs. Taken together, they "state that with increasing content, probability decreases, and *vice versa*; or in other words, the content increases with increasing *im*probability."[141] If growth of knowledge means that we operate with theories of increasing content, Popper concludes, then it must also mean that we operate with theories of decreasing probability (or increasing improbability). "Thus, if our aim is the advancement or growth of knowledge, then a high probability [. . .] cannot possibly be our aim as well."[142] Given that low probability equals a high degree of falsifiability, the aim of science equals a high degree of falsifiability, or testability—or a high informative content.[143]

PROBABILITY

Probability occupies the whole of Chapter 8 of *The Logic of Scientific Discovery*, the longest of the book. And in Chapter 9 Popper applies the tools he has just developed to quantum theory. Furthermore, if we consider the many appendixes, in great part dealing with the theory of probability or its applications to quantum theory, we realize that Popper devoted over half of *The Logic of Scientific Discovery* to probability—not to mention Part 2 of the first volume of the *Postscript to The Logic of Scientific Discovery* and the whole of volume three (on the propensity interpretation of quantum theory). The reason, he says, is that "we still lack a satisfactory, consistent definition of probability; or, what amounts to much the same, we still lack a satisfactory axiomatic system for the calculus of probability."[144] Thus, he confronts two tasks: first, provide new foundations for the calculus of probability; and second, elucidate the relations between probability and experience, which means solving what he calls the problem of decidability of probability statements.[145]

Among the various interpretations of axiomatic systems for the calculus of probability, Popper distinguishes subjective and objective ones. The former treat the degree of probability "as a measure of the feelings of certainty and uncertainty, of belief or doubt, which may be aroused in us by certain assertions or conjectures."[146] In other words, subjective theories interpret probability as a measure of the incompleteness of our knowledge. One of their variants, however, "interprets probability statements not psychologically but *logically*, as assertions about what may be called the 'logical proximity' of statements."[147] Objective interpretations, by contrast, treat "every numerical probability statement as a statement about the *relative frequency* with which an event of a certain kind occurs within a *sequence of occurrence*."[148]

The problem, to repeat, is that of the decidability of probability statements. Just like universal statements, probability statements cannot be verified. But, unlike universal statements, they cannot be falsified either, since "Probability hypotheses *do not rule out anything observable*; probability statements cannot contradict, or be contradicted by, a basic statement; nor can they be contradicted by a conjunction of any finite number of basic statements; and accordingly not by any finite number of observations either."[149] We should therefore "describe them as empirically uninformative, as void of empirical content."[150] Yet, any such conclusion clashes with the successes achieved by physics with predictions obtained from hypothetical estimates of probabilities: "a physicist is usually quite well able to decide whether he may for the time being accept some particular probability hypothesis as 'empirically confirmed,' or whether he ought to reject it as 'practically falsified,' *i.e.*, as useless for purposes of prediction."[151] The problem, thus, is to provide an assessment of probabilistic hypotheses in the face of their unquestionable empirical utility, despite their being logically unfalsifiable (or unverifiable). Just as in the case of Duhem's thesis, Popper's proposed solution is to suggest the integration of logic and methodology: "It is fairly clear that this 'practical falsification' can be obtained only through a *methodological decision* to regard highly improbable events as ruled out—as prohibited."[152] By introducing a methodological rule, Popper transforms the problem of decidability of statistical hypotheses, making them falsifiable.

Frequency theory, initially upheld by Popper, fails to account for single-case probabilities, such as "the probability of throwing five with the next throw of this die is 1/6."[153] This restricts its applicability, whereas subjective interpretations do contemplate singular probabilities. However, Popper needs an objective interpretation, one which enables him to counteract the subjective interpretation of quantum theory.[154] In particular, the reinterpretation of the two-slit experiment advanced in *The Logic of Scientific Discovery* presupposes the calculus of single-case probabilities—thus convincing Popper, by the way, of the reality of propensities.

At first, Popper hoped to solve the problem of single-case probabilities within the frequency interpretation, by defining "The formally singular probability that the event k has the property β—given that k is an element of the sequence α—is, by definition, equal to the probability of the property β within the reference sequence α."[155] As such, the value of the formally singular probability equals that of its objective probability. Later on, he specifies that "the statement 'the probability that the next toss will be heads is one-half' *means* the same as the hypothesis 'the relative frequency of the heads is one-half'; that is to say, the sentence only seems to be singular, but should be properly interpreted as one about a sequence."[156]

In 1957 Popper himself raises an objection to his own proposal[157], which he subsequently reformulates a few times.[158] His conclusion is twofold: on the one hand, "a singular event may have a probability even though it may occur only once; for its probability is a property of its generating conditions";[159] on the other, "we have to visualise the conditions as endowed with a tendency, or disposition, or propensity, to produce sequences whose frequencies are equal to the probabilities; which is precisely what the propensity interpretation asserts."[160]

The introduction of hidden propensities underlying frequencies allows Popper, on the one hand, to highlight the objective feature of probability—as opposed to those who interpret it subjectively, as a measure of the imperfection of our knowledge—and, on the other hand, to account for single-case probabilities. For, whereas the frequency theory attributes probability to a single event merely insofar as it is an element of a sequence with a relative frequency, "the propensity interpretation attaches a probability to a single event as a representative of a *virtual or conceivable sequence* of events, rather than as an element of an actual sequence."[161] The crucial difference between the frequency and the propensity interpretations lies precisely in the status of singular probability statements:

> They play a peripheral role in the frequency theory but a central role in the propensity interpretation which sees, as it were, every single case as the outcome of a propensity, or perhaps of contesting propensities, even though these can be tested only empirically.[162]

In the propensity interpretation, that is, every single case is the outcome of a propensity that can be tested statistically: as opposed to the frequency interpretation, which requires infinite sequences in order to compute probabilities, the propensity theory needs only to grasp the trend—the propensity, that is—of physical phenomena.

The significance of Popper's measure-theoretical approach

> lies in the fact that measure-theoretical probability statements are *singular* probability statements: statements that assert what we may call a *"singular probability."* Yet *from the point of view of physics, a singular probability which "almost entails" a frequency can be best interpreted as a physical propensity.* Thus the mathematical transition from frequency to measure theory corresponds, I suggest, to *a transition from the statistical to the propensity interpretation of objective physical probabilities.*[163]

The concept of propensity, or of a field of propensities—in analogy with the physical concept of field of force—"introduces a dispositional property of singular physical experimental arrangements—that is to say, of singular physical events—in order to explain observable frequencies in sequences of repetitions of these events."[164] Thus, the main argument in favour of the propensity interpretation "is to be found in its power to eliminate from quantum theory certain disturbing elements of an irrational and subjectivist character—elements which are more 'metaphysical' than propensities and, moreover, 'metaphysical' in the bad sense of the word."[165] What Popper is proposing, that is, "is *a new physical hypothesis* [. . .] testable and [. . .] corroborated by certain quantum experiments."[166]

THE PROPENSITY INTERPRETATION OF PROBABILITY

In a lecture delivered in Bristol in 1957 Popper outlines his proposal for an interpretation of the theory of probability in terms of propensities, spelling out its main theses as follows:

1. The solution of the problem of interpreting probability theory is fundamental for the interpretation of quantum theory; for quantum theory is probabilistic theory.
2. The idea of statistical interpretation is correct, but is lacking in clarity.
3. As a consequence of this lack of clarity, the usual interpretation of probability in physics *oscillates* between two extremes: an *objective* purely statistical interpretation and a *subjective* interpretation in terms of our incomplete knowledge, or of the available information.
4. In the orthodox Copenhagen interpretation of quantum theory we find the same oscillation between an objective and a subjective interpretation: *the famous intrusion of the observer into physics.*[167]
5. As opposed to all this, a revised or reformed statistical interpretation is here proposed. It is called the *propensity interpretation of probability.*
6. The propensity interpretation is a purely objective interpretation. It eliminates the oscillations between objective and subjective interpretation, and with it the intrusion of the subject into physics.
7. The idea of propensities is "metaphysical," in exactly the same sense as forces or fields of forces are metaphysical.
8. It is also "metaphysical" in another sense: in the sense of providing a coherent programme for physical research.[168]

Later on, Popper further develops the propensity interpretation of probability, devoting the whole second part of the first volume of the *Postscript to The Logic of Scientific Discovery* to it. At the end of the second volume he shows "that what has stood so long in the way of a conscious acceptance of the propensity interpretation has been the belief in metaphysical determinism";[169] in the third volume, Popper puts to the test the usefulness of the propensity interpretation by applying it to quantum theory and shows that, "with the help of the propensity interpretation, a new metaphysics of physics can be constructed—a new research programme for physics which unifies most of its older programmes and which, in addition, seems to offer possibilities for a unification of the physical and the biological sciences."[170] Finally, in a public lecture delivered in 1988—and later collected in *A World of Propensities*—Popper develops his most philosophically challenging thesis: we live in a world of propensities.

The underlying assumption is that propensities "are not mere possibilities but are physical realities: They are as real as forces, or fields of forces."[171] Just as forces are propensities for setting bodies in motion, so fields of forces are propensities distributed over some region of space, and perhaps changing continuously over this region. That is why "propensities should not be regarded as properties *inherent in an object*, such as a die or a penny, but [. . .] as *inherent in a situation* (of which, of course, the object [is] a part)."[172] In physics, as well as in chemistry, biochemistry, and biology, propensities "are properties of *the whole physical situation* and sometimes even of the particular way in which a situation changes."[173]

We live in a changing world, in which situations—and, with them, possibilities and hence propensities—continuously transform. Our very understanding of the world modifies the conditions in which we happen to live, just as our desires, preferences, hopes, hypotheses, and theories (and false theories as well) do: contrary to the claims of determinism, "indeterminism and free will have become part of the physical and biological sciences."[174]

Past circumstances do not determine the future, but rather determine changing propensities that influence future situations, without determining them univocally. Whereas the past is fixed, the future is objectively not fixed: "The future is open: objectively open"[175]—and the present "can be described as the continuing process of the actualization of propensities; or, more metaphorically, of the freezing or the crystallization of propensities."[176] The world is no longer a causal machine—one big, extremely precise clockwork mechanism, as Descartes would have it—but "it can now be seen as a world of propensities, as an unfolding process of realizing possibilities and of unfolding new possibilities."[177]

It was these tendencies, Popper argues, which gave rise to the appearance of life and the multifaceted unfolding of its various forms. In turn, the evolution of life has provided better conditions, and hence new possibilities and new propensities. In this sense, Popper says, possibilities that have not yet actualized do have a sort of reality. At every moment there are possibilities that try to become actual, but only some of them will succeed. When a given moment nears, most of them would be set to 0, while others would become very small; the remaining ones would become bigger, until they assume the value 1 in the moment they become actual.

The evolution of life moves along the same line: across the centuries, a nearly infinite number of possibilities has developed. At every evolutionary step exclusive choices have annihilated many possibilities, so that only a relatively small number of propensities have become actual. And yet, in a process that has witnessed the fusion of preferences and accidents, the variety of propensities that have actualized is amazing. Something similar, we might conclude, happened to Popper: since he was very young he was always attracted by theoretical problems, problems connected with the growth of scientific knowledge, and particularly the theory of probability. The proposals he advanced are mere accidents, but extremely fruitful ones, which can give rise to many other problems, encourage criticism, and stimulate new reflections.

3   Metaphysics

The idea of freedom is the core of Popper's philosophy and he is determined to make it the ideal background of his very worldview.[1] Beginning with the 1950s, Popper thinks it appropriate to insert his epistemological—as well as political—reflections within a wider "metaphysical" framework.[2] Popper's considerations on realism, indeterminism, World 3, and the self came to constitute a specific and distinct phase in the development of his thought, within which his epistemological and political ideas move back, so to say, to take a secondary place. The central problem became that of cosmology, that is, as Popper declares in the 1959 preface to the revised English edition of *Logik der Forschung*:

> *the problem of understanding the world—including ourselves, and our knowledge, as part of the world.* All science is cosmology, I believe, and for me the interest of philosophy, no less than of science, lies solely in the contributions which it has made to it.[3]

Popper's point of departure is that "in almost every phase of the development of science we are under the sway of metaphysical—that is, untestable—ideas; ideas which do not only determine what problems of explanation we shall choose to attack, but also what kinds of answers we shall consider as fitting or satisfactory or acceptable, and as improvements of, or advances on, earlier answers."[4] They are, indeed, metaphysical research programmes, which remain mostly unconscious in the minds of scientists but shape their judgements and attitudes, thus influencing their assessments and orienting their choices. Popper calls them "metaphysical," "because they result from general views of the structure of the world and, at the same time, from general views of the problem situation in physical cosmology";[5] and he calls them "research programmes," "because they incorporate, together with a view of what the most pressing problems are, a general idea of what a satisfactory solution of these problems would look like."[6]

Despite their being general views (or pictures) of the world, and not empirically testable theories about it, they turn out to be necessary for science, since "they largely determine its problem situations."[7] And they do so not as mere tools, required in order to do research, but as guides for it: they help scientists to decide whether to take a hypothesis seriously or not, whether it is a potential new discovery, and how its acceptance might influence the problem situation. It is perhaps possible, Popper claims, to find "a criterion of demarcation *within metaphysics*, between rationally worthless metaphysical systems, and metaphysical systems that are worth discussing, and worth thinking about."[8] Although we cannot come to a conclusion, whether positive or negative, it is possible to argue for or against a given metaphysical stance, and compare the arguments thus provided. The fact that one of these pictures is worth considering would depend "upon its capacity to provoke rational criticism, and to inspire attempts to supersede it by something better."[9]

REALISM

Popper has always claimed to be a metaphysical realist:[10] to him, being a realist simply means to think, in agreement with common sense, that the world exists independently of men. That is to say, "my own existence will come to an end without the world's coming to an end too."[11] As well as other metaphysical positions, realism is a nontestable conjecture: "realism is neither demonstrable nor refutable."[12] The alternative position, idealism, is equally metaphysical—but both are arguable for, and Popper undertakes the task to show that the weight of arguments that may be advanced in a discussion about them is overwhelmingly in favour of realism.[13]

Perhaps, Popper says, the strongest argument in support of realism consists in a combination of the following two. First, realism is part of common sense: "all the alleged *arguments* against it are not only philosophical in the most derogatory sense of this term, but are at the same time based upon an uncritically accepted part of common sense."[14] Second, "almost all, if not all, physical, chemical, or biological theories imply realism, in the sense that if they are true, realism must also be true":[15] for, the aim of science is to describe or, as far as it is possible, to offer a satisfactory explanation of reality,[16] and the conjectures that we appeal to in order to do that tend to zero in on the truth. In the third place, any arguments in support or against realism have to be formulated in some language, and human language is essentially descriptive (and argumentative): "Rationality, language, description argument, are all about some reality, and they address themselves to an audience. All this presupposes realism."[17] Furthermore, in Popper's eyes "idealism appears absurd,"[18] since it would imply that the world is a product of our mind—that Beethoven's symphonies, Dürer's engravings, or Michelangelo's sculptures do not really exist, but have somehow been created by us: "Denying realism amounts to megalomania (the most widespread occupational disease of the professional philosopher)."[19] Finally,

If realism is true [. . .] then the reason for the impossibility of proving it is obvious. The reason is that our subjective knowledge, even perceptual knowledge, consists of dispositions to act, and is thus a kind of tentative adaptation to reality; and that we are searchers, at best, and at any rate fallible. There is no guarantee against error. At the same time, the whole question of the truth and falsity of our opinions and theories clearly becomes pointless if there is no reality, only dreams or illusions.[20]

These and other arguments persuaded Popper to accept realism "as the only sensible hypothesis—as a conjecture to which no sensible alternative has ever been offered."[21] All epistemological arguments in support of alternative stances, such as positivism, phenomenalism, phenomenology, and so on, are in his view not only mistaken, but—what is worse—they "are the result of the mistaken quest for certainty, or for secure foundations on which to build."[22]

INDETERMINISM

"My dream programme is metaphysical. It is non-testable: it is irrefutable (and irrefutability, we should remember, is not a virtue but a vice). It is based upon the metaphysical (rather than the 'scientific') idea of indeterminism."[23] Popper deems indeterminism as crucial, and, from the 1950s on, keeps going back to it.[24]

As it is customary, Popper begins by considering the contrasting view of Laplacian determinism, whose intuitive idea he captures with an image: "the world is like a motion-picture film: the picture or still which is just being projected is *the present*. Those parts of the film which have already been shown constitute *the past*. And those which have not yet been shown constitute *the future*."[25] In such a film past and future coexist, both being equally determined. For, according to this view, "the state of the universe at any moment of time, future or past, is completely determined if its state, its situation, is given at some moment, for example, the present moment."[26] However confirmed by *prima facie* deterministic successful physical theories, such a view is for Popper "the most solid and serious difficulty in the way of an account of, and a defence of, human freedom, creativity, and responsibility."[27]

The idea of determinism is of religious origin, in connection with the ideas of divine omnipotence and omniscience. There are other forms of determinism, though: the first is scientific determinism, which can be seen as "the result of replacing the idea of God by the idea of nature, and the idea of divine law by that of natural law."[28] As in the case of God, nature is attributed omnipotent and omniscient features—but, as opposed to God, nature can be investigated in order to discover the laws governing it. Once discovered, these laws would allow us to predict the future by means of purely rational methods and on the basis of the data describing the present: "the fundamental idea underlying 'scientific' determinism is that the structure of the world is such that every future event can in principle be rationally calculated in advance, if only we know the laws of nature, and the present or past state of the world."[29] A third version of the deterministic doctrine is metaphysical determinism, according to which "all events in this world are fixed, or unalterable, or predetermined."[30] It does not assert that they are known to anybody, or somehow predictable by scientific means. Only, it asserts that the future, just like the past, cannot be changed.

Like other metaphysical instances, determinism is not testable, and hence not refutable. However, it is possible to criticize it and weight the arguments for or against it. The strongest ones are those supporting scientific determinism: if these fail, little is left in support of metaphysical determinism, too.

One of the fundamental assumptions in support of determinism is "the popular idea of causality,"[31] according to which every event is "caused" by another, and hence it must be determined in advance by the events that constitute its cause. The intuitive commonsense ideas of causality and event are mainly qualitative and do not ask for the quantitative precision that determinism requires. Therefore, if the idea that any event must have a cause may well be regarded as correct, even if only to some extent, it cannot be taken in support of determinism.[32] Moreover, the belief that the world is predetermined may lead to some intuitively absurd results: it would not be possible, for instance, to create a work such as one of Mozart's symphonies simply by studying his brain physically and physiologically.[33] In Popper's eyes, however, the decisive argument against determinism is from the asymmetry between the past and the future: for if it is true that the past, as such, is completely determined by what has happened, the same cannot be said about the future, as determinism would have it. It is common sense itself that testifies it:

> All our lives, all our activities, are occupied by attempts to affect the future. Clearly, we believe that what will happen in the future is largely determined by the past or the present, for all our present rational actions are attempts to influence, or to determine, the future. [. . .] But just as clearly, we do look upon the future as not yet completely fixed; in contrast to the past which is closed, as it were, the future is still open to influence; it is not yet completely determined.[34]

Popper reinforces the argument from the asymmetry between the past and the future by appealing to Einstein's special theory of relativity, according to which for every observer—or, for every local inertial system—there exists an absolute past and an absolute future, separated by contemporaneity. In physical terms, then, the asymmetry between the past and the future "is established by the fact that from any place in the 'past,' a physical causal chain (for example, a light signal) can reach any place in the 'future'; but from no place in the future can such an effect be exerted on any place in the past."[35] As a consequence, "the future becomes '*open*' to us in the sense that it cannot be fully predicted *by us*, while the past is 'closed.'"[36] In other words, according to the special theory of relativity, "the past is that region which can, in principle, be known; and the future is that region which, although influenced by the present, is always 'open': it is not only unknown, but in principle not fully knowable."[37]

A third argument in support of indeterminism, besides the approximate character of scientific knowledge and the asymmetry of the past and the future, is the fact that "*we cannot predict, scientifically, results which we shall obtain in the course of the growth of our own knowledge.*"[38] For, if we could predict today what theory will occur to us in a month's time, then the theory would occur to us today, of course. Although it is impossible to use this very argument against metaphysical determinism, it is a very good argument against scientific determinism, and therefore against any pretence to found deterministic conceptions upon the success of science.[39]

The possibility of making predictions from within the world being ruled out, we are left with the option that the world, seen from without, might be predetermined, possibly by some God. It is the hypothesis advanced by metaphysical determinism and picked up by Einstein in a conversation with Popper in Princeton, in 1950: in his view, the world was a four-dimensional block universe, as unchanging as Parmenides' three-dimensional block universe.[40] Popper opposes this view with two arguments: first, nothing, in our experience, justifies the adoption of a metaphysics *à la* Parmenides;[41] and second, if the universe is assumed to be predetermined and four-dimensional, a number of consequences follow which are "hard to accept."[42] For, if the future is causally determined by the past, we can view it as contained in the past, and therefore it becomes redundant, superfluous. Moreover, that would also imply that the time we experience is an illusion and that the arrow of time is purely subjective, that is, there is no privileged direction in which time flows. Finally, "if we were experiencing successive shots of an unchanging world, then one thing, at least, was genuinely changing in this world: our conscious experience. [. . .] And since we are part of the world, there would thus be change in the world."[43]

As in the case of realism, these are not conclusive arguments. They suffice, however, for Popper to conclude that—contrary to what determinists believe—the future is open, and that "all clocks are clouds, to some considerable degree—even the most precise of clocks."[44] It is, once again, a personal choice—the choice for human freedom and creativity—that involves a precise ethical dimension.

However, if the option for indeterminism is necessary, it is not enough.[45] For indeterminism

> [i]s by itself not enough to make room for human freedom: it is not enough to make human freedom understandable. To do this, I assert,

we need more. We need in addition at least the causal openness of what I am going to call World 1 towards World 2, as well as the causal openness of World 2 towards World 3, and vice versa.[46]

THE THEORY OF THE THREE WORLDS

His insistence on the objective aspect of knowledge brings Popper to elaborate, around the mid 1960s, a theory of the objective mind, or the theory of World 3.[47] Indeed, he distinguishes three worlds or universes:

> first, the world of physical objects or of physical states; secondly, the world of states of consciousness, or of mental states, or perhaps of behavioural dispositions to act; and thirdly, the world of *objective contents of thought*, especially of scientific and poetic thought and works of art.[48]

In other words, World 1 is made of objects that we normally experience: tables, trees, animals, planets, and so on. It is objective, because its inhabitants can be experienced also by others; and it is autonomous, because their existence does not depend upon us. World 2 comprises all our psychological experiences: states of mind, feelings, wishes, memories, emotions. It is subjective, because a person cannot experience the mental states of another person; and it is not autonomous, because the existence of mental states depends on the existence of the mind that experiences them. Other objects do not fall in either of these two worlds: words and propositions, for instance, books and symphonies, laws, numbers and triangles—besides problems, theories, and arguments. These are immaterial objects (though they might be at times contained into material objects, such as in the case of books and music scores), as opposed to the inhabitants of World 1; and whereas World 2 inhabitants are subjective, they are objective; furthermore, as to autonomy, despite that they are products of the human mind, once created, they have consequences that their own creators had not foreseen nor could predict. As a consequence, such objects belong to a different world, World 3:[49] this is inhabited by the set of products of all our cultural activities and comprises all human works from the point of view of their logical and objective content.[50]

Inhabitants of Popper's World 3 not only exist, but mutually interact. We create an object of World 3 when we take an object of World 2 and articulate it by some means—language, a movie, a musical piece—that others can understand. In so doing, we can treat thought as if it were an object: we can put it in front of us, study it, modify it, disassemble it and see its components, and possibly correct it. Most importantly, since World 3 objects can be false, we can realize that something is wrong, or that something can be done so that things work properly. Not only, then, can we contribute to World 3, but we can help to improve both our contributions and others'.

Besides the direct interaction between World 2 and World 3, World 1 and World 2 directly interact as well, whereas World 1 and World 3 interact only indirectly.[51] When we feel pain we move away from its cause, and our physical movement is the result of the interaction between World 1 and World 2. When we read a text and accept (or reject) what it says, our mental processes result from an interaction between World 2 and World 3. But when we read a book—a cookbook, for example—and, following its content, bake a cake, our actions are the product of the interaction of World 1, World 2, and World 3: World 1 and World 3 can interact indirectly by way of the human mind, represented by World 2. For, while baking a cake, a mind (World 2) can appeal to a theory (World 3) in order to modify certain objects (World 1). But World 1 and World 2 can exert a critical control over World 3 objects as well: for, after tasting a cake, we can decide whether we should modify the recipe, and in what ways.

Popper's World 3 has strong similarities with Plato's theory of Forms or Ideas, and with Hegel's theory of the Objective Spirit (or Absolute Spirit), though it is closer to Bolzano's theory of a universe of statements in themselves and truths in themselves, or to Frege's universe of objective contents of thought.[52] Plato was the first to suggest the existence of the intelligible world, beyond the purely sensible one, and Popper's World 3 is indeed somewhat similar to Plato's hyperuranic realm.[53] There are important differences, though: whereas Plato's Forms or Ideas are divine, unchanging, and eternal, the inhabitants of Popper's World 3 are human products that autonomously evolve through time; from this point of view, they are similar to a bird's nest, a beaver's dam, a spider's web, or a walking path in the forest.[54] Furthermore, whereas Plato (as Frege after him) introduced the theory of Forms or Ideas in the attempt to provide humans with ultimate explanations by appealing to essences, that is, to provide the metaphysical scaffolding required for scientific statements to aspire to certainty, Popper does not regard objective knowledge as certain, and his World 3 objects may well turn out to be false.[55] "Plato's theory is a theory of descent or degeneration—a theory of our fall—while mine is a theory of evolutionary ascent towards world 3."[56]

There are significant differences with Hegel's theory of Objective (or Absolute) Spirit as well, which is in continuous evolution: although such Spirit is a human product, "the individual creative element, the relation of give-and-take between a man and his work [. . .] degenerates into the doctrine that the great man is something like a medium in which the Spirit of the Epoch expresses itself."[57] Moreover, the evolutionary pattern of Hegelian dialectics—the similarities with Popper's model of the growth of knowledge by trials and errors notwithstanding—assigns contradictions a positive role, whereas contradictions, in Popper's view, are only errors to be eliminated. Finally, Popper does not accept Hegel's doctrine "that the

Spirit is not only conscious, but a self":[58] according to him, World 3 has no similarity whatsoever with any human conscience, and its objects are utterly different from thoughts in the subjective sense.

Furthermore, although Bolzano's statements in themselves and truths in themselves clearly belong to World 3, he did not manage to see clearly their relationship with the rest of the world: by contrast, Popper seeks to do that, comparing the status and autonomy of World 3 with those of animal objects, and showing how World 3 grows from the higher functions of human language.[59] In addition, Bolzano's postulated world, however as real as World 3, does not possess the kind of reality of the physical world, nor was he able to provide any explanation of the relationship between the two worlds.[60] As to Frege's *dritte Reich*[61]—and the Stoics' *lektón*[62]—it consists of concepts and true and false statements, but no reference is made to arguments. Finally, Frege's "third realm" lacks the epistemological (and particularly Darwinian) outlook Popper assigns to World 3, by claiming that its best inhabitants are scientific theories.[63]

SUBJECTIVE KNOWLEDGE AND OBJECTIVE KNOWLEDGE

Popper distinguishes two kinds of knowledge: *"knowledge or thought in the subjective sense,* consisting of a state of mind or of consciousness or a disposition to behave or to react; and [. . .] *knowledge or thought in an objective sense,* consisting of problems, theories, and arguments as such."[64] In his eyes, traditional epistemology has misunderstood scientific knowledge by deeming it an object of World 2. In so doing, it focused on the wrong problems, such as the specifications of the conditions by which a person is justified in regarding his beliefs as true: "traditional epistemology with its concentration on the second world,[65] or on knowledge in the subjective sense, is irrelevant to the study of scientific knowledge."[66] By contrast, Popper argues that

> what is relevant for epistemology is the study of scientific problems and problem situations, of scientific conjectures [. . .], of scientific discussions, of critical arguments, and of the role played by evidence in arguments; and therefore of scientific journals and books, and of experiments and their evaluation in scientific arguments.[67]

On the one hand, the study of the World 3 of objective knowledge has a decisive importance for epistemology;[68] on the other, "an objectivist epistemology which studies the third world can help to throw an immense amount of light upon the second world of subjective consciousness, especially upon the subjective thought processes of scientists; but *the converse is not true."*[69] World 3 "is a natural product of the human mind, comparable to a spider's web";[70] it is largely autonomous, even though we constantly act upon it and are acted upon by it, by way of a strong feedback process; finally, objective knowledge grows through this intense interaction, and there is a close analogy between the growth of knowledge and biological growth, that is, the evolution of plants and animals. Popper calls "the approach from the side of the products—the theories and the arguments—the 'objective' approach or the 'third world' approach. And [. . .] the behaviourist, the psychological, and the sociological approach to scientific knowledge the 'subjective' approach or the 'second-world' approach."[71] Whereas the subjective approach is causal, that is, it proceeds from the causes to the determination of effects, the objective approach "starts from effects rather than causes,"[72] and it is the ordinary approach in all sciences. Therefore, the problem is that "of the relation between knowledge in the objective sense and knowledge in the subjective sense."[73] More than that: "we cannot understand the first thing about subjective knowledge except by studying the growth of objective knowledge and the give and take between the two kinds of knowledge."[74]

In Popper's view, the philosopher should not be concerned with the subjective aspect of knowledge—that is, the dispositions that cause individuals to uphold a theory with greater or lesser strength—rather, with its objective aspect, which consists "of the logical content of our theories, conjectures, guesses."[75] If subjective knowledge presupposes the existence of a knowing subject, "Knowledge in the objective sense is *knowledge without a knower*: it is *knowledge without a knowing subject*,"[76] since it disregards the personal dispositions and inclinations of individuals and assesses a theory independently of them.

Objective knowledge belongs by definition to World 3: indeed, it constitutes its biologically most important part, that with greater repercussions on World 1. Objective knowledge consists of hypotheses, theories, but also unsolved problems and arguments in support or against a particular theory. Accordingly, the growth of objective knowledge is part of the growth of World 3 and determines its evolutions. As a consequence, Popper's approach to World 3, as well as that to the problems of the relationship between body and mind, and the growth of knowledge, bears a biological orientation.

THE BODY-MIND PROBLEM

In 1977, in collaboration with John C. Eccles (Nobel Prize for Medicine, 1963), Popper publishes *The Self and Its Brain*. By way of autonomous essays and long conversations the two authors intend to defend the idea of psychophysical interaction. Popper, in particular, in the context of a tightly woven criticism of materialism, advances his own version of "downward causation" (the World 3 of theories, the World 2 of the mind influence, and the World 1 of the body and its behaviour):[77] "Darwinian theory, together with the fact that conscious processes exist, lead beyond physicalism,"[78] and together with the thesis of the existence of an evolved conscious it leads to interactionism.

Shortly before his death, Popper revised and gathered in *Knowledge and the Body-Mind Problem* a series of lectures delivered at Emory University in May 1969. In a nutshell, the thesis of the volume is that "in order to understand the relationship between the body and the mind, we must first recognize the existence of objective knowledge as an objective and autonomous product of the human mind, and, in particular, the ways in which we use such knowledge as a control system for critical problem-solving."[79] Although based on lectures given some twenty-five years before, the book contains Popper's most complete and extensive treatment of the subject.

The main thesis is that

> We cannot understand world 2, that is, the world inhabited by our own mental states, without understanding that its main function is to *produce* world 3 objects, and to be *acted upon* by world 3 objects. For world 2 interacts not only with world 1, as Descartes thought, but also with world 3; and world 3 objects can act upon world 1 only through world 2, which functions as an intermediary.[80]

The body-mind (or brain-mind) problem grows from the attempt to explain the interaction between physical and mental states, or physical and mental processes—that is, to understand whether and how our World 2 thought processes are bound up with brain events in World 1.[81] There have been four main attempts at a solution: *immaterialism, pure mentalism*, or *spiritualism*: the denial of the existence of the World 1 of physical objects (Berkeley, Mach); *pure physicalism*, or *philosophical behaviourism*: the denial of the existence of the World 2 of mental states or events (a view common to certain materialists, physicalists, and philosophical behaviourists, or philosophers upholding the identity of brain and mind); *psychophysical parallelism*, or *body-mind parallelism*: the assertion of a thoroughgoing parallelism between World 2 mental states and World 1 brain events (Geulincx, Spinoza, Malebranche, and Leibniz); *body-mind interactionism*: the assertion that World 2 mental states can interact with World 1 physical states (Descartes).[82]

Popper's position is that "a brain-mind parallelism is almost bound to exist *up to a point*."[83] However, if it is true that certain reflexes, such as blinking when seeing a suddenly approaching object, are of a more or less parallelistic character, "the thesis of a *complete* psychophysical parallelism [. . .] is a mistake."[84] In its stead, Popper proposes a form of *psychophysical interactionism*.

The classic problem—"Descartes' problem"—is the following: "how can it be that such things as states of mind—volitions, feelings, expectations—influence or control the physical movements of our limbs? And [. . .] how can it be that the physical states of an organism may influence its mental states?"[85] That is to say, in Popper's own terminology, it is the problem to explain the mutual influence between World 1 and World 2. However, the fundamental problem can be described as the problem "of the influence of the *universe of abstract meanings* upon human behaviour (and thereby upon the physical universe)."[86] By "universe of abstract meanings" Popper means to refer to "such diverse things as promises, aims, and various kinds of rules, such as rules of grammar, or of polite behaviour, or of logic, or of chess, or of counterpoint; also such things as scientific publications (and other publications); appeals to our sense of justice or generosity; or to our artistic appreciation; and so on, almost *ad infinitum*."[87] It is, in other words, the problem of the influence of World 3 upon World 1: Popper calls it "Compton's problem."

Every acceptable solution to Descartes' or Compton's problem must satisfy what Popper calls "*Compton's postulate of freedom*: the solution must explain freedom; and it must also explain how freedom is not just chance but, rather, the result of a subtle interplay between *something almost random or haphazard*, and *something like a restrictive or selective control* such as an aim or a standard—though certainly not a cast-iron control."[88] Such a postulate restricts the acceptable solutions to the two problems "by demanding that they should conform to the idea of combining freedom and control, and also to the idea of 'plastic control,' as I shall call it in contradistinction to a 'cast-iron' [i.e., deterministic] control.'"[89] It is, once again, the idea of human freedom—the idea that underlies the whole of Popper's philosophical reflection:

> the problem of the body-mind relationship [. . .] includes the problem of human freedom, which in every respect, including politics, is a fundamental problem; and it includes the problem of man's position in the physical world, the physical cosmos.[90]

Besides making explicit the contrast between subjective and objective thought, the theory of the three worlds allows Popper to overcome the body-mind dualism in favour of an interactionist dualism that rejects both reductionist monism and Cartesian dualism.[91] As Descartes, he advances a dualist perspective—his is not a dualism of mutually interacting substances, though, but rather of "*two kinds of interacting states* (or events), physio-chemical and mental ones."[92] Popper articulates his position into five theses:

1. Full consciousness is anchored in world 3—that is, it is closely linked with the world of human knowledge and of theories. [. . .]
2. The self, or the ego, is impossible without the intuitive understanding of certain world 3 theories and, indeed, without intuitively taking these theories for granted. [. . .] Or to put it in another way, the self, or the ego, is the result of achieving a view of ourselves from outside, and thus of placing ourselves into an objective structure. Such a view is possible only with the help of a descriptive language.

3. Descartes' problem of the location of full consciousness or the thinking self is far from nonsensical. My conjecture is that the interaction of the self with the brain is located in the speech centre. [...]
4. The self, or full consciousness, is exercising a plastic control over some of our movements which, if so controlled, are human actions. Many expressive movements are not consciously controlled, and so are many movements which have been so well learned as to have sunk into the level of unconscious control.
5. In the hierarchy of controls, the self is not the highest control centre, since it is, in its turn, plastically controlled by world 3 theories. But this control is, like all plastic controls, of the give-and-take, or feedback, type. That is, we can—and we do—change the controlling world 3 theories.[93]

The new structures that emerge as unforeseeable products of the evolutionary process always interact with the basic structures of the physical states from which they emerge;[94] the control system interacts with the controlled system; mental states interact with psychological states; and World 3 interacts with World 2 and, through it, with World 1.[95]

EVOLUTIONARY EPISTEMOLOGY

As a proposed solution of the problem of the growth of knowledge, Popper advances a simplified "tetradic schema" of the method by conjectures and refutations:

$$P_1 \rightarrow TT \rightarrow EE \rightarrow P_2$$

P_1 is the problem—whether practical or theoretical—from which we start; TT is a tentative theory we advance in order to solve this problem; EE is the process of error elimination, that is, the correction and refinement of the theory by critical tests; P_2 is the problem—or, most likely, the problems that emerge from the previous phase of critical discussion of our proposed theory. Usually, these latter problems are not deliberately created by those advancing the theory, but rather "they emerge autonomously from the field of new relationships which we cannot help bringing into existence with every action, however little we intend to do so."[96]

As a matter of fact, highlights Popper, "*knowledge starts from problems and ends with problems* (so far as it ever ends)":[97] it is problems that characterize the relationship between the living and the world:

The growth of knowledge—or the learning process—is not a repetitive or a cumulative process but one of error-elimination. It is Darwinian selection, rather than Lamarckian instruction.[98]

Popper's theory of the objective mind assigns a crucial importance to language: only once it is linguistically formulated can an object belonging to World 2 be conveyed; only when it is expressed into a common language, that is, can an individual's mental state become something other from the individual—"detached" from him, so to say[99]—and become public, intersubjectively criticizable.[100] More than that: if, on the one hand, language allows for criticism, on the other it makes it necessary—for human language, having developed the higher functions of language that govern description and argumentation, introduced the possibility of telling the false, that is precluded to animals.

Karl Bühler, Popper's teacher at the University of Vienna, distinguished three functions of language: two lower ones (shared by both animals and humans) and a higher one (that characterizes human language). The expressive function, by which speakers express their emotions or thoughts; the signal (or stimulative, or release) function, that serves to stimulate or release certain reactions in the hearer; and the descriptive function, which presupposes the two previous ones and describes a certain state of affairs by way of statements that can be true or false (it is with the descriptive function that criteria of truth and falsity are introduced as well). To these, Popper adds the argumentative or explanatory function, which adds to the previous functions the ability to present and compare arguments or explanations in connection with certain definite questions or problems.[101]

With the development of the argumentative function of language, criticism becomes the main tool for the growth of knowledge, and logic can be regarded as the *organon* of criticism.[102] "The autonomous world of the higher functions of language becomes the world of science,"[103] and the tetradic schema

> [b]ecomes the schema of the growth of knowledge through error-elimination by way of systematic *rational criticism*. It becomes the schema of the search for truth and content by means of rational discussion. It describes the way in which we lift ourselves by our bootstraps. It gives a rational description of evolutionary emergence, and of our *self-transcendence by means of selection and rational criticism*.[104]

The formation of human theories—that is, of objective knowledge—Popper argues, is

> something like a *mutation outside our skin* or, as it is called, an "exosomatic mutation." Theories are in this respect (but not in all respects) like instruments, for instruments are like exosomatic organs. Instead of growing better eyes, we grow binoculars and spectacles. Instead of growing better ears, we grow microphones, loudspeakers, and hearing aids. And instead of growing faster legs, we grow motor cars.[105]

Animals and plants are problem solvers:[106] their anatomy and behaviour embody provisional, tentative solutions that are the biological correspondents of theories advanced by humans to solve a given problem. Just like spider webs,[107] they are exosomatic products or tools that influence World 1, contributing to its change and development.

Traditional epistemology, Popper concludes, by regarding our sensory perceptions as the "data" upon which theories are to be built by some inductive process, can be described as pre-Darwinian:

> It fails to take account of the fact that the alleged data are in fact adaptive reactions, and therefore interpretations which incorporate theories and prejudices and which, like theories, are impregnated with conjectural expectations; that there can be no pure perception, no pure datum; exactly as there can be no pure observational language, since all languages are impregnated with theories and myths. [. . .] This consideration of the fact that theories or expectations are built into our very sense organs shows that the epistemology of induction breaks down even before taking its first step.[108]

Life and scientific research advance along the same line: from old problems to the discovery of new and undreamed of ones. Against what he calls "epistemological expressionism"—the subjective approach according to which knowledge is a relation between the subjective mind and the known object, and the products of the human mind are regarded as mere utterances or expressions of mental states—Popper argues that

> everything depends upon the give-and-take between ourselves and our work; upon the product which we contribute to the third world, and upon that constant feed-back that can be amplified by conscious self-criticism. The incredible thing about life, evolution, and mental growth, is just this method of give-and-take, this interaction between our actions and their results by which we constantly transcend ourselves, our talents, our gifts. This self-transcendence is the most striking and important fact of all life and all evolution, and especially of human evolution.[109]

Just like animals, we appeal to imaginative criticism to transcend the environment we happen to live in. We criticize universality, or the structural necessity of what might appear as "given"; we struggle to find, devise, and construct new, critical solutions; we try to spot, discover, and doubt our own prejudices and all we take for granted. This, Popper says, is "how we lift ourselves by our bootstraps out of the morass of our ignorance; how we throw a rope into the air and then swarm up it—if it gets any purchase, however precarious, on any little twig."[110]

Popper looks at the growth of knowledge "as if it were a struggle for survival between competing theories."[111] Just as in natural selection, those theories are eliminated that offer the wrong solution of a given problem. As opposed to natural selection, though, in the field of knowledge "*we can let our objective theories die in our stead.*"[112] What makes our efforts different from those of any animals is that the rope we throw may get a purchase in the World 3 of critical discussion: this is what makes it possible for us to reject some of the theories competing to solve our problem.

In this perspective, life becomes a struggle against the obstacles that, along the way, oppose the self-assertion of individuals and the realization of their values. The aim is to modify the world surrounding us in order to make it a better environment in which we can live and prosper:[113] "life is problem-solving and discovery—the discovery of new facts, of new possibilities, by way of trying out possibilities conceived in our imagination."[114] Such trying out takes place almost exclusively in World 3, through the attempts to represent, by the theories that inhabit it, the objects of both World 1 and World 2. It is not only the environment that selects and changes us, but it is we who select and change the environment as well. "We do not mould or 'instruct' [the third world] by expressing in it the state of our mind, nor does it instruct us. Both we ourselves and the third world grow through mutual struggle and selection."[115]

4 Popper and Kuhn
Clashing Metaphysics

Having offered a picture of Popper's antifoundationalism, I wish now to contrast it with a different picture: Thomas Kuhn's view of science. For, by Popper's own admission, "Professor Kuhn's criticism of my views about science is the most interesting one I have so far come across."[1]

Orthodoxy, especially in light of the enormous impact of Kuhn's ideas, presents us with a picture of a sharp and massive break, of a true revolution: Kuhn is viewed as a philosopher whose main achievement is to have undermined a whole philosophical tradition, that of Logical Positivism. I think this is wrong: for, from many and often fundamental points of view, Kuhn did not manage to break entirely with the preceding philosophical tradition; his works are laden with principles belonging to that very empirical philosophy he was determined to reject. Furthermore, only a partial challenge of positivism and empiricism can actually account for the genesis of Kuhn's philosophical perspective—incommensurability, the notion of progress, the rejection of the concepts of truth and verisimilitude, and the very thesis of "world change" (one of the theses deemed most radical and characteristic of Kuhn's philosophical stance) are all consequences of the empiricist elements that his philosophy retains. For sure, Kuhn played a major role in the "historical turn" that marked philosophy of science in the last third of the past century, thus contributing to the radical shift of focus from logic and language analysis to a more historically informed approach, concerned with the dynamics of theory change and conceptual change. Appearances to the contrary notwithstanding, however, the implicit presuppositions and the stated principles of Kuhn's philosophy are not very different from those of the logical positivists or logical empiricists he saw himself to be distancing from.

Far from spelling the "official demise"[2] of Logical Positivism, then, Kuhn's philosophy is its natural continuation (which is, perhaps, the reason why Carnap welcomed his book so favourably[3]). Like Copernicus, who, while dealing the first fatal blow to the Aristotelian-Ptolemaic worldview, was also irrevocably soaked in that very same way of thinking, so Kuhn can be regarded as the last exponent of the philosophical tradition he was determined to reject. In *The Copernican Revolution* Kuhn wrote that

Copernicus is like the middle point of a bend in a road: from one point of view it is the last point of one stretch of the road; from the other, it is the beginning of the next. In this sense, Kuhn is the last of the neopositivists and, at the same time, the first of their successors. In particular, Kuhn's later "linguistic turn" marks a clear step back into inductivism and justificationism, completely disregarding Popper's philosophical revolution and turning Kuhn's followers into foundationalist philosophers.

KUHN ON TRUTH AND REALISM

In the first edition of *The Structure of Scientific Revolutions* (1962) Kuhn hardly refers to the concept of truth: he has no need of it, not even in order to characterize and explain progress.[4] For, in the closing pages of the book, Kuhn writes:

> The developmental process described in this essay has been a process of evolution *from* primitive beginnings—a process whose successive stages are characterized by an increasingly detailed and refined understanding of nature. But nothing that has been or will be said makes it a process of evolution *toward* anything.[5]

He then urges us to give up the concept itself in order to get rid of some of the problems which have afflicted the history of Western thought:

> We are all deeply accustomed to seeing science as the one enterprise that draws constantly nearer to some goal set by nature in advance. But need there be any such goal? Can we not account for both science's existence and its success in terms of evolution from the community's state of knowledge at any given time? Does it really help to imagine that there is some one full, objective true account of nature and that the proper measure of scientific achievement is the extent to which it brings us closer to that ultimate goal? If we can learn to substitute evolution-from-what-we-know for evolution-toward-what-we-wish-to-know, a number of vexing problems may vanish in the process. Somewhere in this maze, for example, must lie the problem of induction.[6]

In the 1969 "Postscript" to the second edition of the book he introduces two arguments against the notion of truth implicit in the traditional view of progress as increasing verisimilitude:

> A scientific theory is usually felt to be better than its predecessors not only in the sense that it is a better instrument for discovering and solving puzzles but also because it is somehow a better representation of what nature is really like. One often hears that successive theories grow

ever closer to, or approximate more and more closely to, the truth. Apparently generalizations like that refer not to the puzzle-solutions and the concrete predictions derived from a theory but rather to its ontology, to the match, that is, between the entities with which the theory populates nature and what is "really there." Perhaps there is some other way of salvaging the notion of "truth" for application to whole theories, but this one will not do.[7]

In the 1980s and 1990s, he relates his rejection of truth to incommensurability.[8] Indeed, as a consequence of Kuhn's later characterization of incommensurability, which is ascribeed both an inevitable and a functional role for the growth of scientific knowledge, there is no need for the notions of "truth" and "approximation to the truth." Kuhn has always opposed the correspondence theory of truth and criticized its applications to the relation between scientific theories and reality: as history can show, he says, there is

no theory-independent way to reconstruct phrases like "really there"; the notion of a match between the ontology of a theory and its "real" counterpart in nature [. . .] seems to me illusive in principle. Besides, as a historian, I am impressed with the implausibility of the view. I do not doubt, for example, that Newton's mechanics improves on Aristotle's and Einstein's improves on Newton's as instruments for puzzle-solving. But I can see in their succession *no coherent direction of ontological development.*[9]

The basic idea of traditional epistemology, a correspondence theory of truth that assesses beliefs on the grounds of their ability to reflect the world, independently of the mind, cannot account for the change of these very beliefs, according to Kuhn.[10] Therefore, it must be rejected and replaced with a weaker conception,[11] internal to the paradigm—or lexicon[12]—itself. For if a statement can be properly said to be true or false within the context of a given lexicon, the system of categories embedded in the lexicon cannot be, *per se*, truth or false.[13] By relinquishing the correspondence theory of truth Kuhn rejects the idea that the system of categories of a theory may reflect the world-in-itself, independently of theory. We may speak of truth only within the context of a given lexicon, that is, we may only assess the assertions stated within a given lexical context: "lexicons are not [. . .] the sorts of things that can be true or false":[14] their logical status is that of words' meaning in general, that is, of a convention we can only justify in a pragmatic way. Truth is internal to lexicon in the sense that its use is restricted to assessing claims made within the context of the lexicon: truth claims in one lexicon are not relevant for those made in another, nor can truth be applied to a lexicon itself.[15]

In other words, Kuhn decidedly rejects the idea that the structure which constitutes the theory might reflect the way the world is, independently of theory. The lexicon embodies a linguistic convention that marks the distance between the reality described by a theory and the theory describing it in different ways:

> Experience and description are possible only with the described and describer separated, and the lexical structure which marks the separation can do so in different ways, each resulting in a different, though never wholly different, form of life. Some ways are better suited to some purposes, some to others. But none is to be accepted as true or rejected as false; none gives privileged access to a real, as against an invented, world. The ways of being-in-the-world which a lexicon provides are not candidates for true/false.[16]

Lexicons are assessed on the basis of their ability to serve a particular function, not to reflect reality.[17] To quote Kuhn's own words once again:

> what replaces the one big mind-independent world about which scientists were once said to discover the truth is the variety of niches within which the practitioners of these various specialties practice their trade. Those niches, which both create and are created by the conceptual and instrumental tools with which their inhabitants practice upon them, are as solid, real, resistant to arbitrary change as the external world was once said to be. But, unlike the so-called external world, they are not independent of mind and culture, and they do not sum to a single coherent whole of which we and the practitioners of all the individual scientific specialties are inhabitants.[18]

In its stead, he argues, a weaker concept of truth can be retained: we can talk about truth only within the context of a given lexicon. Indeed, within such a lexical structure a claim may be properly said to be true or false, but that does not hold for the system of categories embedded in the lexicon, which is not itself capable of being true or false.[19] The lexicon has the status of a linguistic convention which may be judged only on the basis of how well it serves a particular purpose, rather than how well it reflects reality. As a result, though Kuhn assumes the existence of an independent reality throughout his work, his position involves idealistic leanings.[20]

Just as Kuhn—in Wittgenstein's footsteps—understands errors only in terms of preestablished rules,[21] so can he conceive truth only within the context of a given linguistic framework. The idea that lexicons (or paradigms) are not and cannot be true or false *per se* is but a variant of justificationism:[22] it is the idea that truth is grounded on the solidarity of beliefs within a given scientific community, an immediate consequence of Kuhn's highlighting of the communitarian character of science. Indeed, it is one of the most important elements Kuhn shares with Positivism. For positivists as well placed particular emphasis on community: they regarded communal collaboration as important for the production and justification of scientific knowledge, which they in turn regarded as important for the unity of science. It is this very emphasis that fuels Kuhn's conception of science as a social institution and his attempt to define scientific knowledge, if not truth itself, in terms of the consensus of belief that is forged among its members. As Feyerabend highlighted, however, "advance of knowledge [. . .] has nothing to do with membership in communities (Wittgenstein notwithstanding)."[23]

THE BACKGROUND OF KUHN'S POSITION

Whereas Popper clearly separated justification and criticism, in Kuhn—as well as in Wittgenstein—justification and criticism remain inextricably combined. That is why he cannot appeal to criticism as alternative to justification and pleads for the description of conceptual frameworks and standards in its stead. In fact, Kuhn's position is rooted both in justificationism and in a particular way of posing problems which is typical of Wittgenstein and his followers.[24] Taken together, these two closely interwoven aspects work together and reinforce one another, forcing the compartmentalization of knowledge and the limitation of rationality.

From David Hume (1711–1776) onwards, it has been asserted that there are two kinds of inference: deductive inference, which defines logic; and inductive inference, which defines the natural sciences.[25] The two apply, so to say, to different fields, and must not be confused: the problem of induction is simply dissolved once we learn not to apply the standards of deductive logic to judge inductive inference. Once we realize that the two principles cannot be unified, the task of the philosopher is simply that of describing and clarifying the standards of deductive and of inductive reasoning. Most positivists, while maintaining the unity of the sciences, accepted this "methodological" division. Wittgenstein extended this approach: each discipline, or field, or "language game," or "form of life" is alleged to have its own standards, or principles, or "logic," which need not conform to or be reducible to any other standards and which, again, is the special task of the philosopher to describe and clarify—not the least to judge, defend, or criticize.[26] There is no arguing or judging among disciplines: criticism, evaluation, and explanation would no longer be proper philosophical aims. Knowledge is essentially divided, and description is all that remains to the philosopher. All he can do is to describe the logics, grammars, or first principles of the various kinds of discourse, and the many sorts of language games and forms of life in which they are embedded. Philosophical critique is no longer of content, but of criteria application: as Paul Feyerabend put it, all that is left are "consolations for the specialist."[27]

Regardless of Kuhn's own intentions, his philosophy drastically impoverishes the reasons, aim, and scope of philosophical critique. It is more concerned with the acceptance of ideas, rather than with their content; it legitimizes existing structures[28] and overlooks the aims of those working within it, particularly the growth of knowledge.[29] Confrontation is often banned, criticism discouraged.[30] Philosophical and scientific values of truth and rationality are replaced with commitment to the dominant tradition and consensus within a community of experts.[31]

By contrast, as Mark Notturno rightly notices,[32] Popper's solution to the problem of induction offers an account of how scientific knowledge can be objective and rational without being certain and without grounding itself upon expert opinion, consensus, and the solidarity of belief. By regarding criticism, not description, as the alternative to justification, it offers an account in which truth is the key element.

LONDON 1965: KUHN VERSUS POPPER

In July 1965, the International Colloquium in the Philosophy of Science, held at Bedford College in Regent's Park, London, provided the occasion for a confrontation between Popper and Kuhn.[33]

Kuhn begins his paper by highlighting the close resemblance between his own position and Popper's:[34] both reject the idea that science progresses by accumulation; both emphasize the revolutionary process by which an older theory is overthrown and replaced by a new one (but whereas Kuhn adds that the two are "incompatible,"[35] Popper does not stress this feature and often speaks of theories that survive as special cases of new ones, of which they constitute a good approximation); both, in particular, emphasize "the intimate and inevitable entanglement of scientific observation with scientific theory" and "are correspondingly sceptical of efforts to produce any neutral observation language";[36] finally, though by Kuhn's own admission the list is not exhaustive, both insist that scientists aim at providing explanations of observed phenomena and do so in terms of real objects. Apparently, Popper and Kuhn dealt with the very same data, but got different *Gestalten* from them—so much that Kuhn was tempted to title his paper "Logic of Discovery or Psychology of Research: A *Gestalt* Switch?"[37]

The differences between them were equally noteworthy, though. In the first place, whilst Popper says that scientists presuppose their theories and then test them, hence advancing the idea that knowledge grows through a continuous overthrowing of ideas, Kuhn highlights that scientists first assume a "constellation"[38] of theories shared by the scientific community, and then put to test not that constellation, but their very ability and ingenuity to solve the puzzles they face during their research activity.[39] Revolutions, as Kuhn understands them, are very rare episodes in the history of science. That is why Popper, as Kuhn sees him, "characterized the entire scientific enterprise in terms that apply only to its occasional revolutionary parts. [. . .] a careful look at the scientific enterprise suggests that it is normal science, in which [Popper's] sort of testing does not occur, rather than extraordinary science which most nearly distinguishes science from other enterprises."[40]

Secondly, whereas Popper sees science as a special case of the process through which we learn from our errors, in Kuhn's eyes learning from our errors makes sense only against the background of a set of accepted rules and procedures that can be employed to identify a single failure in applying them. Therefore, according to Kuhn, learning from our errors takes place only during periods of normal science. Thirdly, rather than employing terms such as "refutation," Kuhn speaks of a paradigm that is no longer able to sustain a puzzle-solving tradition. When a sufficiently high number of scientists become convinced of this inability, they decide to transfer their commitment to another paradigm (if any) that is able to keep the promises the old one proved unable to keep.

Finally, after affirming the impossibility of defining a satisfying notion of verisimilitude, and therefore of speaking of progress in terms of ever better approximations to the truth,[41] Kuhn underlines that *philosophical explanation* of scientific progress must, "in the final analysis, be psychological or sociological. It must, that is, be a description of a value system, an ideology, together with an analysis of the institutions through which that system is transmitted and enforced."[42]

Whilst Kuhn and Popper often seem to be upholding closely similar positions, a decidedly conspicuous disagreement marks their views of criticism. For Popper the testing process undergone by a theory is a special case of the unceasing critical discussion of foundations which alone can warrant the rational character of science without bringing it down into dogmatism, while for Kuhn history suggests that "it is precisely the abandonment of critical discourse that marks the transition to a science."[43] The clash involves the very basic assumptions of falsificationism. Popper insists on the rational nature of science, marked by the openness of mind of its practitioners that allows it to grow and progress. Extreme flexibility of thought and creative boldness are balanced by a relentless demand for the refutability of our hypotheses. Scientists, according to Popper, should make an effort to refute their own theories rather than seek confirmations of them. The hallmark of intellectual honesty is in stating in advance under what conditions we would be ready to give up our theories. Without considering irrelevant questions of meaning, Popper is firmly convinced that scientific theories progress towards an ever-better correspondence with reality. By contrast, Kuhn seems to be drawing a picture of a scientific community like a closed society, formed by closed-minded people, bounded by and committed to certain procedural models—paradigms—that guide their theoretical and experimental activity.[44] Practitioners of a certain discipline attempt to frame nature in the bounds given by the paradigm. Revolutions are rather occasional events, usually the outcome of the scientists' inability to assimilate and analyze facts in the way the paradigm suggested. They are processes akin to religious conversions and commit the members of a scientific community to a new system of theories, practices, and methods. Radical, but more often subtle meaning changes of key theoretical terms see to it that scientists bound to the new paradigm manage only partially to communicate with those supporting the old one. Although some of these changes can lead to actual improvements in the level of understanding of nature, Kuhn does not speak of approximation to the truth.

For Popper science is an essentially critical enterprise, and therefore it is revolutionary in perpetuity.[45] Science is always revolutionary because it is thought in evolution, that is, critical thought. The discovery of something is always a discovery *against* something else, because, as in the case of Christopher Columbus, it collides with a constellation of established prejudices.[46] The creative scientist does not seek an easy consensus, but gives rise to a frank dissent, even if difficult to handle, since "only in the change of a system [. . .] [is it] clearly shown the character of a science that draws teachings from reality, from experience."[47] Thus Copernicus went beyond the tradition affirmed by Ptolemy, Newton went beyond Galileo and Kepler, and Einstein beyond Newton. Overthrows of established ideas (the so-called "revolutions of ideas") are not exceptional episodes, but constitute the usual condition of scientific activity: science grows as a *revolution in permanence*.

For his part, Popper always stressed "the need for some dogmatism: the dogmatic scientist has an important role to play. If we give in to criticism too easily, we shall never find out where the real power of our theories lies."[48] But this does not seem to be the kind of dogmatism Kuhn advocates: "He believes in the domination of a ruling dogma over considerable periods; and he does not believe that the method of science is, normally, that of bold conjectures and criticism."[49] Moreover, Kuhn's arguments are logical ones: "Kuhn suggests that the rationality of science presupposes the acceptance of a common framework. He suggests that rationality *depends* upon something like a common language and a common set of assumptions. He suggests that rational discussion, and rational criticism, is only possible when we have agreed on fundamentals."[50] This is for Popper the thesis of relativism, and it is a logical one: "the myth of the framework," as he labels it, "a logical and philosophical mistake."[51] Admittedly, in every moment we are prisoners caught in the framework of our language, theories, past experiences, and expectations—but we are prisoners "in a Pickwickian sense: if we try, we can break out of our framework at any time.[52] Admittedly, we shall find ourselves again in a framework, but it will be a better and roomier one; and we can at any moment break out of it again."[53] Critical discussion, in other words, is always possible, and the contrary thesis (i.e., the incommensurability thesis, the idea that different frameworks are like mutually untranslatable languages) is a dangerous dogma— "the central bulwark of irrationalism."[54] The "myth of the framework" exaggerates a difficulty into an impossibility: however difficult, there is nothing more fruitful than the clash between different cultures of ideas. Denying this possibility is a mistake, since authentic progress springs from it. Incommensurability, in other words, however often taken for granted as a problem, reveals itself rather as a solution, an all too easy way out of problems: instead of confronting them, we deem them insurmountable, label them incommensurable, and set them aside.

Popper believes in "absolute" or "objective" truth, in Tarski's sense: for him scientific knowledge can be regarded as knowledge without a knowing subject.[55] He believes in scientific progress as a progress towards truth, that is, the growth of knowledge. Kuhn is sceptical on this point, and Popper calls him a "relativist":[56] it is "the deepest issue" that divides them.[57] Even if we do not have any method of discovering scientific theories, of ascertaining the truth of a scientific hypothesis, or whether a hypothesis is "probable," or "probably true," we can improve our knowledge through a confrontation (and a clash) between different theories, or hypotheses.

REASON AND THE SEARCH FOR TRUTH

The core difference between Popper and Kuhn is not about the possibility of falsification, incommensurability, or the existence of normal science. It is about the role of truth, the value of criticism, and the nature of the bond that unites scientists into a community. Popper and Kuhn agree that there is no objective criterion for truth, but Kuhn takes this to mean that truth plays no role at all in theory appraisal and theory choice, while Popper maintains that truth plays the role of a regulative idea.[58] As a consequence, Kuhn characterizes the bond uniting scientists in terms of shared beliefs (underlain by shared practices): since it is not possible to prove the truth of such beliefs, scientists cannot help but commit themselves to them uncritically. Popper, by contrast, characterizes this bond in terms of the search for truth, believing that only truth and the critical attitude enable a scientific community to be an open society.

In hindsight, the radical challenge of *The Structure of Scientific Revolutions* was not to rationality, but to realism: Kuhn's thrust was actually directed not so much against the rationality of theory appraisal and theory choice, as against the epistemic, or truthlike, character of the theories so chosen, since it is not possible to say that they are better approximations to the truth, i.e., reality. In so doing, however, Kuhn conflated the concept of truth with the criterion of truth, thus claiming that it makes no sense to speak of truth *per se* in the absence of a decisional procedure to determine it.

The crisis of Logical Positivism betrays the deeper crisis of foundationalism, a philosophical approach that spans the whole Western philosophical tradition. Kuhn was unable to provide a viable alternative. His efforts to this end notwithstanding, the later phase of his philosophy—the so-called "linguistic turn" of the 1980s and 1990s—and the unsuccessful attempts to conclude his last book betray a failure.

Kuhn's contribution to the philosophy of science grew from his attempts to do a history of science from a theoretical point of view. In so doing, he triggered a revolution. He said that revolutions are often started by outsiders, and his own career—that of a physicist who became a historian for philosophical purposes[59]—represents a particularly interesting case. However, as Kuhn himself stressed, revolutions are not often total revisions of the system of beliefs from which they originate. Again, Kuhn's case is an exemplary one: the revolution he triggered retained many aspects of the logical empiricist tradition against which he wished to react. In order to find a viable response to the twentieth-century crisis of foundationalism, we have to acknowledge Kuhn's results, realize the failure of his approach and move on, away from him.

Popper's own answer to the crisis of foundationalism, as outlined in the preceding chapters, is completely different. For, far from being a mere "transitional figure,"[60] or "boundary philosopher,"[61] between Logical Positivism and the "new philosophy of science," Popper's philosophy paves a new way between dogmatism and relativism, thus providing a sound reaction to the crisis of foundationalism that characterizes the twentieth century.

As Notturno once again emphasizes, the fundamental epistemological fact of the past century was the crisis of foundationalism. Its fundamental epistemological problem has been how to respond to it. Some philosophers have concluded that scientific knowledge is unjustified and hence irrational. Others—indeed, the majority—have opted for retaining the idea that scientific knowledge is justified, but weakened either the idea that truth is correspondence with reality or the idea that justification shows that a statement is true. These responses retain the foundationalist theory of rationality, according to which it is irrational to accept a belief that has not been justified and obligatory to accept one that has. Wittgenstein's idea that science is grounded in a form of life; Carnap's idea that it is grounded in a linguistic framework; Kuhn's idea that it is grounded in the acceptance of a paradigm, or lexicon; Rorty's idea that it is grounded in the solidarity of community (thus leading to a downright relativism)—each of these is a return to Hume, since the fundamental idea of each of them is that our knowledge is grounded, and must be grounded if we are to regard it as rational.[62] By contrast, "[f]aced with the impossibility of certain truth, critical rationalists retain their interest in truth, and discard not only certainty but all its surrogates."[63]

The clear realization of the failure of all foundationalist approaches presents us with the possibility of a choice. Popper's original model for the growth of knowledge by conjectures and refutations is a viable alternative. Instead of trying to build on the ruins of the collapsed edifice of the positivist research programme, as Kuhn himself did (and as his heirs and those of the Vienna Circle keep trying to do, only working on additional layers of ruins), we are presented with the project for a new edifice. Popper's original proposal for a rationality without foundation, his view of truth as the regulative idea of science, and his understanding of reason as the negative faculty of relentless criticism can constitute a viable and satisfactory response to the collapse of any foundationalist approach.

RATIONALITY WITHOUT FOUNDATIONS

The appropriate standards of criticism are not those that appeal to justification, but those that appeal to truth.[64] As soon as we give up the idea that we can justify our theories, the epistemic problem becomes the problem not of what to believe, but of how to criticize our beliefs. While traditional philosophy regards the justification of beliefs as the goal, and the critical method a way of reaching it,[65] for Popper the goal is truth, and it is the method of investigation itself that is rational. Rationality, in other words, is not a property of knowledge, but a task for researchers: being rational means nothing but being willing to appeal to reason and arguments, as opposed to violence and force, to resolve our disputes. It requires no foundation, no grounding upon expert opinion or consensus. It needs only critical dialogue.[66]

The history of Western philosophy is marked by attempts to provide our knowledge with a bedrock foundation. The twentieth century opened with a profound crisis in both mathematics and physics: the foundations of these disciplines were shaken and so was the entire edifice of science. More than ever, perhaps, philosophers witnessed the crisis of foundationalism and tried to respond to it. None of these responses frees itself from the ghost of foundationalism. Popper offers an alternative.

The crisis of foundationalism has no implication for truth. It does not show that truth does not exist, and it does not show that it is solidarity and consensus. It shows that—as Socrates beautifully said—we are living in the twilight zone between knowledge and ignorance, where the views that we hold may be true, but where we are unable to know that they are.[67] The failure of foundationalism is not the failure of epistemology: however difficult to reach, Socrates' twilight zone between knowledge and ignorance is an ideal well worth holding on to.

Popper's proposal is that of a *rationality without foundations*. Like the foundationalists and the irrationalists he believed that we must, in science, ultimately make some sort of unjustified decision. Indeed, he saw the rational attitude as a moral obligation and as a clear option against violence, deeming both dogmatism and voluntarist irrationalism to be irresponsible.[68]

Traditional epistemology sets itself the task of finding what it calls the foundations of our knowledge, i.e., a (restricted) body of knowledge which is absolutely certain and from which we can obtain the rest of it in a simple and straightforward fashion. On the contrary, to use the words of one of Paul Feyerabend's most beautiful works,

> any decision against methods creating certainty will be at the same time a decision against the acceptance of foundations of knowledge [. . .]; *it will be a decision in favour of a form of knowledge that possesses no foundation*. And it will therefore also be a decision to leave the traditional path of epistemology and to build up knowledge in an entirely new fashion.[69]

5 The Ethical Nature of Popper's Understanding of Rationality

As Sextus Empiricus reports,

> [. . .] the Skeptics were in hopes of gaining quietude by means of a decision [. . .], and being unable to effect this they suspended judgment; and they found that quietude, as if by chance, followed upon their suspense [. . .]. We do not, however, suppose that the Skeptic is wholly untroubled [. . .].[1]

Life is "skeptical"—in the etymological meaning of *sképsis*, the Greek term for *research*—as Xenophanes taught, since

> The gods did not reveal, from the beginning,
> All things to the mortals; but in the course of time,
> Through seeking they may get to know things better.[2]

With his view of life as a never-ending process of problem-solving, Popper recalled and gave new meaning to the idea of a philosopher who lived two and a half thousand years ago, according to which "all is but a woven web of guesses."[3] Rationality is opinion and action in accordance to reason. However, what this amounts to remains disputed by philosophers, and the theory of rationality grows from such disagreement.

JUSTIFICATION AND CRITICISM

As we have seen, in Popper's view science is defined by the rules of its method: "The game of science is, in principle, without end. He who decides one day that scientific statements do not call for any further test, and that they can be regarded as finally verified, retires from the game."[4] Not retiring means acknowledging the historical dimension of knowledge and stressing the role of individual researchers, who are regarded as responsible people charged with decisions. This holds not only for the theory of knowledge and science. It holds for social sciences with equal force. That is why

Popper grounded his "theory of democratic control" upon "the decision, or upon the adoption of the proposal, to avoid and to resist tyranny."[5] Once again, "decision" is the key word. We need methodological decisions both in democracy and in the scientific enterprise. We could actually replace some words in the passage quoted here and read it as follows: "The game of [democracy] is, in principle, without end. He who decides one day that [rulers and their acts] do not call for any further test, and that they can be regarded as finally [approved], retires from the game." That is, he retires from democracy. We must come to grips with our fallible nature, both in the natural and the social sciences. And in both fields we have to face the problem of the limits, either of our critical ability or of sovereignty. We have to avoid immunizing strategies and practice democratic vigilance. In this sense, Popper's philosophy effected a structural transformation in philosophy as a whole.

"How do we know? How do we justify our beliefs?": as Popper taught us, all questions of this kind beg authoritarian answers, such as "the Bible," "the leader," "the intellect," "sense experience," and the like. Notwithstanding the efforts to free these allegedly indisputable authorities from various difficulties in the course of the Western philosophical tradition, they all proved to be not only inadequate justifications, but also fallible and questionable in themselves. In the same line, "Who should rule?" is the wrong way of posing the problem. Again, such a question begs an authoritarian answer, such as "the best," "the wisest," "the people," "the proletariat," "the chosen ones," or "the master race." Therefore, Popper reformulates the problem in the following way: "How can we best arrange our political institutions so as to get rid of bad rulers, or at least restrict the amount of harm they can do?"[6] It is a radical modification of perspective, for it contains the recognition that there is no best kind of supreme political authority for all situations, and reopens the door to a rational approach in political philosophy: it enables one to be a political rationalist and a kind of democrat without committing oneself to the belief that any majority is right.

Most importantly, what holds true for political philosophy applies perhaps even more significantly to philosophy in general. All proposed sources of knowledge are fallible and epistemologically insufficient—but all are welcome, given that they can be criticized. In William Bartley's words:

> The authoritarian structuring of philosophy's fundamental epistemological question can be remedied by making a shift comparable to the one suggested for political philosophy. We may not only reject (as did the critical rationalists) the demand for rational proofs of our rational standards. We may go further, and *also* abandon the demand that everything *except* the standards be proved or justified by appealing to the authority of the standards, or by some other means. [. . .] *Nothing gets justified; everything gets criticized.* Instead of positing infallible intellectual authorities to justify and guarantee positions, one may build a

philosophical program for counteracting intellectual error. One may create an ecological niche for rationality.[7]

Once again, we could rephrase our problem, switching a few words and putting it into another context: "How can our intellectual life and institutions be arranged so as to expose our beliefs, conjectures, policies, positions, sources of ideas, traditions, and the like—whether or not they are justifiable—to maximum criticism, in order to counteract and eliminate as much intellectual error as possible?"[8]

By sharply separating the concepts of justification and criticism, and by explicitly eliminating the notion of justification from that of criticism, Bartley intends to escape the dilemma of ultimate commitment to an authority, thus eliminating what he thought were the last remnants of a fideistic attitude and of positivism (that is, jutificationism) in Popper's philosophy.[9] His comprehensively critical rationalism, later renamed *pancritical rationalism*, proposes itself as a new philosophical programme, a nonjustificational philosophy of criticism comprising a new conception of rationality. Within such a framework, being rational means being willing to entertain any position and hold anything in it (including the most fundamental standards, goals, and decisions) as open to criticism, without resorting to any authority, faith, or irrational commitment. Any position may be held rationally provided that it remains open to criticism and survives severe tests. There is no theoretical limit to criticizability, that is, to rationality.[10]

IRRATIONAL FAITH IN REASON

The attitude of rational argument, Popper says, cannot be grounded on rational argument. Critical rationalism ultimately relies upon an "irrational faith in reason," a consequence of a moral decision in favour of rationalism:

> whoever adopts the rationalist attitude does so because he has adopted, consciously or unconsciously, some proposal, or decision, or belief, or behaviour; an adoption which may be called "irrational." [. . .] we may describe it as an irrational *faith in reason*.[11]

Bartley regarded this as a too generous concession to irrationalism, a dangerous chink in the armour of critical rationalists (the *tu quoque* argument, as he called it: the most effective weapon in the armory of irrationalism—something we owe to ourselves to answer). He did not want to share any leap of faith, even to reason, and therefore rejected it. In so doing, however, he rejected the moral decision at the root of Popper's philosophy as well. In Bartley's eyes, Popper's assertion about faith in reason attempts to justify critical rationalism, whilst Popper said he needed no justification at all: when Popper identifies a positive basis for his rationalist attitude by means of a moral decision, Bartley argues, he actually falls back on the very justificationist attitude he tried to eliminate. Bartley does not want to have anything to do with basic decisions or moral presuppositions, and therefore proposes to eliminate any kind of justificationism by holding the attitude itself open to criticism.

For Popper, by contrast, rationalism requires a complementary notion of reasonableness, that is, "an attitude of readiness to listen to critical arguments and to learn from experience."[12] This is the moral core of Popper's fallibilism: having realized how little we know, we must not only be fully prepared to correct our mistakes, but we are also required to have doubts about our knowledge. The process of doubting must be a conscious attitude of openness to criticism, which has an individual and a social aspect. On the one hand, each participant in the game of critical discussion is required to be prepared to listen to criticism, to be able to accept criticism, to practice self-criticism, and to engage in mutual criticism with others. On the other, once a subjective attitude or moral stance has been adopted by the individuals, reasoning must be conceived as a social process of intersubjective confrontation.

Reasoning is engaging in communication with others; it requires non-epistemic values of social conduct. Central among these is the moral imperative to take others and their arguments seriously, i.e., to respect them, to be ready not only to allow differences to exist, but to try to learn from them. Popper chooses reason primarily because of its beneficial consequences: rationalism comprises a set of principles that are both epistemological and ethical, and sets the social and political rules for the human cooperation necessary for the acquisition of knowledge. Popper felt it as a very concrete issue, and understood it as a personal choice: "I felt that where moral problems come in, one must not be abstract. [. . .] This was for me, not for my students, but this was for me a kind of faith, or decision, or something like that; for me only. I even didn't advocate it."[13]

By contrast, Bartley's critique seems entirely drawn from logic, thus disregarding the profound ethical nature of Popper's choice.[14] And even if we take pancritical rationalism to emphasize an attitude, possessing itself an ethical dimension (in fact, Bartley arrived at it while trying to solve existential problems relating to religion and ultimate commitment) we cannot help but see how pancriticism seems to be centred more on overcoming the opponent, rather than aiming at avoiding violence.

Popper understands his own moral decision as a choice, as a kind of categorical imperative valued for its own sake, not as a premise to an argument. It is not a commitment to a theory, or to anything that could be accepted or rejected as true or false. His critical rationalism is "fundamentally, an attitude,"[15] not a theory—that is, a disposition, a readiness to listen to each other's critical arguments, to search for one's own mistakes, and to learn from them, following the best argument in a critical debate.

Therefore, it cannot be replaced by a theory of rationality. A theory of rationality is a proposed solution to the problem of rationality. Like any theory, it can be true or false. On the other hand, an attitude is neither true nor false. But again, even if we read Bartley's pancritical rationalism as an attitude—namely, the attitude that requires that everything should remain open to criticism, including this very attitude itself—we have to notice that they are two different attitudes, confronting two different problems. On the one hand, we have the growth of knowledge and the improvement of society and its institutions; on the other, the will to demolish the argument of those relativists, sceptics, and fideists who reproach the rationalist with the *tu quoque* argument.

In fact, when we argue in favour of or against something, we have already adopted or accepted a rational attitude, no matter how tentatively. Rationality is just a word to describe the correct way of finding out what is going on by using unlimited criticism. It has nothing to do with discovering thoughts or assuming stances; it does not allow us to follow a procedure which would be "right" and would lead us to the desired results. Reason is the negative faculty of relentless criticism (*ratio negativa*). Do we wish to become more rational or less rational? Once we accept it this way, we are already rational to some degree. After asking, we can go one way or the other, but the very ability to ask it tells us we are rational to some degree.

> My rationalism is not dogmatic. I fully admit that I cannot rationally prove it. I frankly confess that I choose rationalism because I hate vio-lence, and I do not deceive myself into believing that this hatred has any rational grounds. Or, to put it another way, my rationalism is not self-contained, but rests on an irrational faith in the attitude of reason-ableness: I do not see that we can go beyond this. One could say, per-haps, that my irrational faith in equal and reciprocal rights to convince others and be convinced by them is faith in human reason; or simply, that I believe in man.[16]

By the expression "irrational faith in reason" Popper designated a moral decision that can be supported by argument—a sensible openness, in the effort to take arguments seriously:

> irrationalism will use reason too, but without any feeling of obligation; it will use it or discard it as it pleases. But I believe that the only at-titude which I can consider to be morally right is one which recognizes that we owe it to other men to treat them and ourselves as rational.[17]

PRACTICAL REASON

In Chapter 24 of *The Open Society and Its Enemies* Popper recognizes that an "irrational commitment to rationalism" of the participants is the precondition to dialogue. Bartley aimed at showing that no such irrational commitment is necessary: dialogue itself contains all the rules. However, this does not solve the problem of encouraging people to debate and accept criticism. Epistemology may set norms for the rational appraisal of theories, but these are embedded in a wider set of values guiding social interaction and communication. Epistemology comprises, and indeed requires, certain nonepistemic ethical norms, such as honesty and toleration, without which epistemic norms would be unworkable. Theories of knowledge may therefore be both instrumentally and inherently political.

However, Popper's model of rationality in science already presumes the existence of a particular form of community and society characterized by free critical discussion. Popper's theory of the growth of scientific knowledge is embedded in a worldview that implicitly values particular social arrangements and particular human capacities. His proposal that rationality consists of critical problem-solving presupposes a minimal prior consensus on certain values, ends, and interests. Epistemology must therefore suggest the social and political preconditions for the successful application of its epistemic norms: to complete his theory of scientific rationality, Popper's task becomes that of developing and refining his underlying social and political theory.

According to Hacohen, even though Popper rightly praised multiculturalism and culture clash as productive of great intellectual advances, and as the setting in which liberty grows, he "underestimated the difficulty of creating a situation that would make such a dialogue possible. Culture clash under conditions of unequal power does not always create dialogue, or advance cosmopolitanism. It may result in oppression."[18] In Hacohen's eyes, in other words, Popper's irrational commitment to reason—the very precondition to dialogue, that is—is insufficient, since for an ideal speech situation power must be neutralized, so that it has minimal influence on dialogue: "Only in a deliberative democracy (or the Open Society) are intersubjective criticism and politics proximate. We do not yet live in such a democracy, and if we ever come close, there will be communities that do not accept intersubjective criticism."[19] Though liberal communicative ideals and Popper's rationality criterion tell us where we should head to, Hacohen concludes, they provide no guidelines for evaluating political proposals.

My own sense is that Popper sees the rational attitude as a moral obligation and as a clear option against violence, deeming both dogmatism and voluntarist irrationalism to be irresponsible. It is an attitude of dialogue, accepted with the full awareness of the difficulty of such a task.[20] Those who accept it do so because, consciously or unconsciously, they have adopted a proposal, or decision, like an "irrational faith in reason": a minimal concession to irrationalism which cannot be determined by argument, even though we can argue in its favour—by explaining its consequences, for instance. It may be assisted by arguments, and we can assess the potential consequences of the decision. It is not, however, "determined" by such consequences, and remains the responsibility of the individual.[21]

Our choice is open, and Popper sees his own as a moral article of faith:

The choice before us is not simply an intellectual affair, or a matter of taste. It is a moral decision [. . .]. For the question whether we adopt some more or less radical form of irrationalism, or whether we adopt that minimum concession to irrationalism which I have termed "critical rationalism," will deeply affect our whole attitude towards other men, and towards the problem of social life.[22]

Dialogue is a value to Popper. He views life as a continuous process of problem-solving: problems arise together with life, and there are problems only when there are values.[23] When violence replaces peaceful debate, reason must give in, and we have to fight to establish the minimal conditions for critical exchange.[24] We do not have to try to make people critically minded. We have no right to force them to offer or accept criticism, nor to learn to participate effectively in a critical discussion: it is their right to refuse to do so. All we can (and have to) do is to try to help them become critically minded if and when they request that. Popper's methodological proposals reflect both hope that rational debate will lead to improvement, and the conviction that criticism entails responsibility. Such responsibility can be encouraged, but not forced. It is the best we can do in order to avoid any kind of dogma or authority: such a general moral requirement for social science is arguably one of Popper's most distinctive achievements.

"REASON ALSO IS CHOICE"

Already in *Die beiden Grundprobleme der Erkenntnistheorie*, but more clearly in *The Logic of Scientific Discovery*,[25] Popper appeals both to logical falsifiability and to the methodological decision to carry on with criticism, thus revealing the bent for history of those researchers who have left the "idol" of certainty behind.[26]

Popper would have never given up his moral decision, since his whole philosophy of science and of society presupposes this moral background, which constitutes his solution to the problem of rationality. The problem of rationality is perhaps the most important of philosophical problems and, in a sense, the core of philosophy itself. It concerns the choice of one's principles and values, says Popper; it concerns the choice of one's lifestyle, says Agassi.[27] Rationality is a part of our way of life, and that goes alike for the rationalist and the irrationalist. What we can decide is whether we like it or not, whether we wish to drive it to its limit and expand it, or take it as it comes and be content, or even try to avoid and repress it and then succeed to some measure.

The solution to the problem of rationality is the very starting point of every philosophical approach, the very choice of one's lifestyle. Popper's approach to philosophy is his solution to the problem of rationality: his whole life is the very embodiment of his understanding of rationality and his solution to its fundamental problem.[28]

Popper showed that there are at least two possible forms of life, which are connected with two forms of knowledge. The choice between them involves one's personal responsibility, and must therefore be made individually, as he himself did. In John Watkins' forceful words:

> Determinism and inductivism, although not bound logically together, are natural coalition partners; for, of all extant epistemologies, it is inductivism that most readily furnishes a causal account of belief-formation. There is likewise [. . .] a natural coalition between indeterminism and falsificationism (whereby scientific knowledge is seen as growing through conjectures and refutations). There seems to me no doubt as to which pair of doctrines offers the more cheerful picture. The first depicts man as an induction machine nudged along by external pressures, and deprived of all initiative and spontaneity. The second gives him the *Spielraum* to originate ideas and try them out. Learning about the world means, on the first view, being conditioned by it; on the second view, it means adventuring within it.[29]

We are confronted with a decision which will affect our whole approach to philosophy and life in general. And since the situation cannot be resolved by any proof but only on the basis of our demands and preferences, we have to give up the desire for certainty, the wish to escape our responsibilities. "Reason also is choice," wrote John Milton:[30] the very act of reasoning implies that of a choice. We are called upon to decide what we like better. Our choice in favour of reason does not guarantee that we will be successful in our search for the truth, nor in the discovery of our errors. It does not even necessarily lead to the avoidance of violence.[31] Why be rational, then? Maybe just because we do not want to give up, or voluntarily limit, the use of our intellect. We are free to make the best use of our freedom. It is up to us to leave us the freedom to arrange our lives in the manner we find most congenial. Our choice has therefore an inescapable ethical nature—in Feyerabend's words: "epistemology, or the structure of the knowledge we accept, is grounded upon an ethical decision."[32]

Notes

NOTES TO THE INTRODUCTION

1. *LSD*, p. 111. A similar passage can be found in *BGE*, p. 136: "The empirical basis of objective science is *nothing absolute*. Science does not rest on a bedrock. Its towering edifice, an amazingly bold structure of theories, rises over a swamp. The foundations are piers going down into the swamp from above. They do not reach a natural base, but go only as deep as is necessary to carry the structure. One does not stop driving them down because one reached firm ground. Rather, one resolves to be satisfied with their firmness, hoping they will carry the structure. (If the structure proves too heavy, and begins tottering, it sometimes does not help to drive the piers further down. It may be necessary to have a new building, which must be constructed on the ruins of the collapsed structure's piers). [...] *The objectivity of science can be bought only at the cost of relativity.* (He who seeks the absolute must seek it in the subjective)".

2. As Schlick (1934) shows, logical positivists grounded science in perceptions and experiences. By contrast, Popper states that: "We must distinguish between, on the one hand, *our subjective experiences or our feelings of conviction*, which can never justify any statement (though they can be made the subject of psychological investigation) and, on the other hand, the *objective logical relations* subsisting among the various systems of scientific statements, and within each of them" (*LSD*, p. 44). Carnap and Neurath made an unsuccessful attempt to overcome the gap between psychology and logic by translating psychological behavior into physicalist language: whether phenomenalist or physicalist, their protocols were logical constructions of experience, "perception statements", records of sense data, translations of observations into formal speech. No real improvement was made by changing mode of expression: they remained attached to the psychological basis (see *BGE*, pp. 429-432 and 438-439, and *LSD*, pp. 95-97).

3. Other labels often used to refer to Popper's philosophy are "fallibilism", for its insistence on the fallible character of human knowledge, and "falsificationism", to highlight the role and function of negative instances in the growth of knowledge. Together with "critical rationalism" – his own favourite – they catch the various facets of Popper's philosophical approach.

4. See *P1*, p. 6; see also *LSD*, p. 32.

5. See Notturno (2000), p. 109. Popper appeals to deductive arguments not because they are conclusive, but because – as opposed to inductive arguments – they allow for criticism and revision.

6. For what follows, see Notturno (2000), ch. 5.

7. See *KBMP*, p. 134. See also Jarvie, Agassi (1973), p. 385, and Bartley (1962, 1984), pp. 233-234, as well as Miller (2006), pp. 50, 112, 141 and 150.
8. See *LSD*, pp. 38, 49-50 and 108-110, as well as *OS2*, pp. 380-381.
9. Agassi (1988), pp. 497-498.
10. In 1937, while reading a paper to a philosophy seminar at Canterbury University College (Christchurch, New Zealand), Popper moves from the theoretical assumptions of *LF* to develop a tightly-knit attack on dialectic: it is the premise of his two major works in political philosophy, *PH* and *OS*, and – I suggest – it is at the same time the bridge that connects, and sheds light upon, the two "branches" of Popper's philosophical reflection. The paper, "What is Dialectic?", was later reprinted in *CR*, ch. 15.
11. *OS2*, p. 225; see also *MF*, pp. XII-XIII.
12. See also *OS1*, pp. 200-201, and *OS2*, pp. 279-280.
13. *LSD*, pp. 280-281; see also *OS2*, pp. 224-225, and *CR*, p. 334: "The scientist is not the man who knows a lot but rather the man who is determined not to give up the search for truth".

NOTES TO CHAPTER 1

1. See also *CR*, pp. 33–37 and 255–256.
2. The discrepancy between the actual development of Popper's thought and the picture he himself depicted was first noted by John Wettersten, see his (1985), (1992), and (2005a), and then by Malachi Hacohen in his intellectual biography of Popper (2000). See also Berkson and Wettersten (1982); ter Hark (2002) and (2004); and Gattei (2004). What follows is based upon these works, particularly Wettersten's and Hacohen's.
3. The supporters of this movement aimed at transforming the very nature of the scholastic system. The initial success notwithstanding, the Dollfuss government (1932–1934) restored the previous situation. The relevance of the school-reform movement for Popper (as well as for Wittgenstein) was first noted by William Bartley: see Bartley (1970) and (1974). See also Hacohen (2000), chap. 3.
4. Bühler, a well-known German psychologist, would later play a major role in the early stages of Popper's intellectual development, Popper being awarded his doctorate under Bühler's supervision. Popper learned from him the psychology of Otto Selz (1881–1943) and the realism of the Würzburg School. By contrast, Schlick represented the latest, neo-positivist philosophy of science: he was at the same time the examiner and the critical target of Popper's doctoral dissertation.
5. See *UQ*, pp. 72–78; for a detailed account, see Hacohen (2000), pp. 132–168.
6. "The Attitude of the Teacher towards the School and the Pupils: Social or Individual Education?" In *FS*, pp. 3–9.
7. Ludwig Wittgenstein was also a subscriber to these journals; just as Popper, he worked in Glöckel's reformed school for a few years.
8. "For a Philosophy of the Idea of Homeland," in *FS*, pp. 10–26.
9. *"Habit" and "Experience of Lawfulness" in Education: A Pedagogic-Structural-Psychologic Essay*, in *FS*, pp. 83–185.
10. See Vaihinger (1911).
11. The second is to be found in *BGE*, vol. 1, without §11; the third in *LF*.
12. In his autobiography Popper traced his rigorous deductivism back to a public lecture Einstein delivered in Vienna in 1919, and his criticism of induction to the years 1926–1928, after which he formulated the criterion of demarcation

(see *UQ*, pp. 31–44; but also *CR*, pp. 33–39 and 255–256; as well as *LSD*, pp. 311–312). In sharp contrast with Popper's own reconstruction, however, in 1927 Popper was still an inductivist in the style of Ernst Mach, despite the fact that just a few years later he expressed bewilderment and dissent at Mach's calling "Fourier's theory of heat conduction a 'model theory of physics' for the curious reason that 'this theory is founded not on a *hypothesis* but on an *observable fact*'" (*LSD*, p. 75).

13. *FS*, p. 87.
14. Ibid., pp. 96–98.
15. Ibid., p. 97: no reference to Husserl, of course.
16. Ibid.
17. Ibid.
18. Ibid., pp. 97–98. "This," Popper adds, "is precisely the mistake into which have fallen, for instance, the psychoanalytic movements (Freud, Adler)" (ibid., p. 98). This passage would support the story told in the autobiography, where Popper says he was interested in distinguishing the (bad) methods of the psychoanalysts from the (good) methods of Einstein (see *UQ*, pp. 36–38). But the problem he was then facing was rather different from the one he later said he was concerned with: the former was put in inductive terms (how can we avoid looking at the facts from preconceived opinions?), whereas the latter was a problem of demarcation, which presumed that we look at the facts from preconceived opinions.
19. See *BGE*, p. xxvi.
20. See Vaihinger (1911), pp. 143–154.
21. *FS*, p. 97.
22. Thus the title of the 1927 thesis: Popper aimed at contrasting learning by mere repetition (*Gewohnheit*) with the complex of features that constitute the dogmatic attitude (for which he coined the term *Gesetzerlebnis*, "experience of law"), especially in children.
23. *FS*, pp. 90–91. Critical thinking is identified with the revision of a conceptual framework in response to contrary experiences. Years later, Popper would regard the mutual exchange between dogmatic and critical thinking as a process by trial and error, fundamental for the growth of knowledge. In 1927 dogmatic thinking is simply an obstacle for the learning process, not a necessary precondition for it: it is confined to psychology, without crossing the boundary into logic, the realm of critical thinking.
24. *FS*, p. 94; see also *CR*, pp. 49–51.
25. *FS*, p. 95.
26. In a section devoted to the general methodological presuppositions of the problem of the status of pedagogy, Popper speaks of "delimitation [*Abgrenzung*] of the empirical part of pedagogy from its non-empirical part" (*FS*, p. 89). Although here Popper employs the very same term—*Abgrenzung*—he would later employ in *LF* to refer to demarcation, its meaning is utterly different: In 1927 Popper was not relying on a deductive method (falsifiability), but rather took inductive method for granted. His problem was to verify the extent to which general theoretical statements are supported by empirical facts, and are consequently valid: he was simply delimiting the kind of science which intends to relate to reality (logic and mathematics excluded, then) through the integral and concrete support of empirical evidence. Moreover, this early discussion of the concept of *Abgrenzung* is a discussion within the field of psychology, while Popper's later discussion of the demarcation problem is framed in terms of a sharp distinction between scientific disciplines (such as physics) and nonscientific ones (such as psychoanalysis and Marxism). In 1927 Popper speaks only of the

boundary between Adler's theory and empirical psychology (Külpe's and Bühler's *Denkpsychologie*) and by no means has in mind the concept of demarcation (or the solution to the problem of demarcation) as we know it from his subsequent works; see *BGE*, pp. xxvii–xxix and 3–5; as well as *LSD*, pp. 34–39. In his 1927 thesis the problem of demarcation is as absent as the problem of induction. This analysis sheds new light also on the profound relationship between the problem of demarcation and the problem of induction. For, as to *Abgrenzung*, both the proposals advanced in the *Hausarbeit* and in *LF* depend on the method employed: the former is based on the inductive method, the latter on the deductive method. Therefore, the two problems are intrinsically interwoven, actually constituting two faces of the same conundrum.

27. Falsified theories—that is, theories that have been proved false—are obviously decidable theories, and would consequently be part of science as well. It is an interesting point, but Popper says nothing about this, because he does not know how to handle falsified theories, yet.

28. *The Problem of Method in Thought Psychology*, in *FS*, pp. 187—260. In this work Popper presents both Schlick's *Allgemeine Erkenntnislehre* (1918, 1925) and Bühler's *Die Krise der Psychologie* (1927), contrasting the two and choosing the former as his critical target.

29. Apart from Vaihinger, Popper also had clearly in mind William Whewell (particularly his *Philosophy of the Inductive Sciences*, 1840), who first introduced this idea into the philosophy of science. See Wettersten (2005b) and Wettersten and Agassi (1991).

30. To the expressive, signal, and descriptive functions of language, highlighted by Bühler, Popper adds the study of the argumentative function, thus showing him to have abandoned the theory of judgement as the adequate theory of the processes of critical thinking, upheld in *"Gewohnheit" und "Gesetzerlebnis" in der Erziehung*.

31. This idea (that is also to appear in the 1931 article) will be important for Popper's later philosophy. He would expand on it in a 1948 lecture on the bucket and searchlight theories of knowledge (now in *OK*, pp. 341–361): science progresses just as Selz's oriented thought-processes, that is, from the statement of problems and the successful attempts to solve them (or, if necessary, from the reformulation of problems so as to make them solvable).

32. *FS*, pp. 245–246; see also pp. 246–247.

33. In the autobiography he says it was very close to Selz's: see *UQ*, p. 76.

34. Most likely, it was due to the influence of the Würzburg School that Popper did not regard Freud's work as part of science (a key element for the early phases of his own intellectual development, as Popper would highlight from the 1950s on). For, in *Die Krise der Psychologie* (1927) Bühler excluded Freud's theory from the scope of science on methodological grounds, and Bühler's work was indeed a central element in Popper's intellectual horizon.

35. Popper employs Carnap's *Abriß der Logistik* (1929); it contains a summary of Russell's results as well as a few brief epistemological applications of them.

36. *Axioms, Definitions and Postulates of Geometry*, in *FS*, pp. 263–390.

37. See Hacohen (2000), pp. 176–177.

38. See Poincaré (1898), (1902), and (1908).

39. See Helmholtz (1866), (1868), and (1876).

40. "How can it be that mathematics, being after all a product of human thought which is independent of experience, is so admirably appropriate to the objects of reality? Is human reason, then, without experience, merely by taking thoughts, able to fathom the propriety of real things [?] In my opinion the

answer to this question is, briefly, this: as far as the laws of mathematics refer to reality, they are not certain; and as far as they are certain, they do not refer to reality" (Einstein 1922, p. 28). See *LSD*, p. 314. See also *CR*, chap. 9.

41. See *LSD*, pp. 82–84.

42. Indeed, it is the idea of truth—of truth as correspondence between the world and the hypotheses that are meant to describe it, as Popper would say beginning in 1935, after his meeting with Tarski—and the key role it plays, which radically distinguishes Popper's position from that of Kuhn (or Wittgenstein, or Rorty), for instance, positions that deprive truth of any role whatsoever within their views, thus dooming to failure any attempt to reconcile their respective positions (such as Lakatos' methodology of scientific research programmes).

43. *CR*, p. 212; this was written in 1946.

44. See also *FS*, p. 307.

45. *FS*, pp. 377–378.

46. For example, Popper accepts Reichenbach's remark that we should not postulate any "universal force," that is, any unnecessary force in order to explain the way in which we perceive the world—a force definitely needed if we want to reconcile, say, Einstein's physics with the statement that space is Euclidean. This might be a glimpse of the later theory of demarcation, but it is not developed at all. Demarcation as a problem is still far from Popper's intellectual horizon.

47. "The Care of Memory from the Point of View of Individual Activity," in *FS*, pp. 27–49. In 1932 Popper published a long bibliographical survey of works in psychology, pedagogy, and education: "Pädagogische Zeitenschriftenschau" ("Survey of Pedagogical Journals"), now in *FS*, pp. 50–79.

48. The title alludes to Arthur Schopenhauer's *The Two Fundamental Problems of Ethics* (1841). Popper took Schopenhauer as a model for his writing style.

49. See *BGE*, p. xxxv.

50. See also the remarks on metaphysics in *BGE*, pp. 3–4, where physics, which has "completely freed itself of the remaining slag of his metaphysical past" through a "stormy process of purification" is contrasted with other sciences "less highly developed (such as biology, psychology and sociology) [which] are imbued with metaphysics much more than physics is, and are still imbued with it."

51. Interestingly, Popper shared this view with Reichenbach: see Reichenbach (1931), particularly pp. 342–344 (this is an article well known to Popper, who referred to it many times in *BGE*, chap. 6).

52. The breakthrough will take place only under the pressure of three different problems: the provisional nature of basic statements; Reichenbach's objection (in his rejoinder, Reichenbach (1933), where he argues that no theory can be refuted) to "Ein Kriterium des empirischen Charakters theoretischer Systeme" ("A Criterion of the Empirical Character of Theoretical Systems", a note published by Popper in *Erkenntnis* in 1933); and the new problem of demarcation, according to which demarcation is not achieved by regarding as scientific those statements which are attributed a truth-value, but by distinguishing empirical theories from other ones.

53. See Reichenbach (1931), p. 342: "we should say that [the principle of induction] serves to decide their [i.e., of scientific theories] degree of probability. The alternatives in science are not truth and falsehood; instead, there is a continuous scale of probability values whose unattainable limits are truth and falsehood." And a few lines below: "probability statements are not even meaningful unless the principle of induction is presupposed. [. . .] Probability statements are not meaningful within a two-valued logic that

requires every statement to be either true or false. [. . .] it turns out that one cannot justify the assertion of probability laws if two-valued logic is regarded as the only criterion for testing our knowledge of reality."

54. See, in particular, Wettersten (1985) and (2005a).
55. See *BGE*, p. 329.
56. Ibid., pp. 3–4.
57. See *BGE*, p. 314: "We can therefore state the laws of nature *as if* the general states of affairs existed. And we are allowed to state them *only* in this way, that is, *only* as provisional assumptions. For this reason—and only in this sense—the laws of nature are 'fictions.' [. . .] As a consequence, the formulation I have already made use of many times, that the laws of nature '*can* not be true,' is not to be understood as saying that it states the *impossibility* that a law of nature can be true. Such a formulation is merely aimed at accounting for the logical impossibility for us to *decide*, on some occasions, the truth of the laws of nature." The difficulty Popper is facing is evident: even when he rejects the idea that the laws of nature are "fictions," he appeals to the idea that they are mere "as if" statements (a formulation that is characteristic of conventionalism). The expression "as if" is Vaihinger's (see the title of his 1911 work), just as the formulation adopted by Popper is his, although he rejects Vaihinger's conventionalism.
58. This is not a satisfactory view and Popper would abandon it immediately in *LF*. However, the reason why he upholds it now—that is, in logic only proved statements can be assigned truth-values, and only formulae with an assigned truth-value can be deemed statements—will not be overcome. In *LF*, however, he will be able to abandon it anyway, because he will no longer be concerned with proof. It will be only after learning Tarski's theory of truth in connection with nonproven statements (in 1935, after the publication of *LF*) that Popper will be confident of overcoming the difficulty represented by his seeming inability to regard an unproved statement as true.
59. See *BGE*, §48.
60. Ibid., p. 327.
61. Ibid., pp. 339–418.
62. As Popper would explain several years later, Tarski's theory allowed him to speak of truth in the absence of a criterion for truth; see *OK*, pp. 319–321.
63. See *CR*, pp. 33–34 and 51–52.
64. See *LSD*, p. 59: "Theories are nets cast to catch what we call 'the world': to rationalize, to explain, and to master it. We endeavour to make the mesh ever finer and finer."
65. See Reichenbach (1931), p. 343: "It is not possible to justify the system of scientific statements simply on the basis of deductive logic together with observational reports; this is our epistemological result."
66. See *LSD*, pp. 312–314; and Reichenbach (1933), p. 427.
67. For an analysis of §11, see Wettersten (1985); (1992), pp. 144–161; and (2005a), pp. 613–617; see also Hacohen (2000), pp. 220–222.
68. We could also think of an original section of *LF*, then drastically reduced to appear in print.
69. Of Jewish origin, Nelson (1882–1927) was a cosmopolite intellectual who preached a universal Kantian ethics. Popper found in Nelson "a model of critical philosophy and progressive politics" (Hacohen 2000, p. 121); "Nelson's cosmopolitanism informed Popper's own, but it was his epistemology that proved essential to Popper's intellectual development. As a student, Nelson discovered the nearly forgotten Kantian philosopher Jakob Friedrich Fries (1773–1843). Fries considered himself Kant's true successor. He formed

a critique of Kant's transcendental proofs in epistemology, ethics, and religion" (ibid., p. 122).
70. Hacohen (2000), pp. 122–123.
71. See also Wittgenstein (1921), 4.1121: "Theory of knowledge is the philosophy of psychology."
72. The change in terminology follows the underlying conceptual change: Popper no longer speaks of singular statements, thus taking into consideration only their logical form, but of basic statements, that is, statements that describe observable states of affairs.
73. See *LSD*, pp. 93–94 and 104–105; see also Hacohen (2000), p. 230.
74. See Neurath (1935).
75. "A Criterion of the Empirical Character of Theoretical Systems," in *LSD*, pp. 312–314.
76. See particularly Neurath (1933) and (1935).
77. See Reichenbach (1933).
78. The editor of *BGE* insists on the actual existence of a complete version of the second volume as well, that has been lost; on this, however, see Wettersten (1985) and (1992), p. 161; as well as Hacohen (2000), pp. 238–242.
79. The division into groups offered here does not follow the (doubtful) chronological order established by Troels Eggers Hansen, but John Wettersten's analysis: see his (1992), pp. 161–164.
80. *BGE*, pp. 353–374 and 389–395, respectively.
81. "Entwurf einer Einführung," "Orientierung," and "Philosophie," ("Outline of an Introduction," "Orientation," and "Philosophy") in *BGE*, pp. 341–346, 383–384, and 385–388, respectively.
82. See *BGE*, pp. 347–349. Such a shift reflects the shift from the study of the logic of scientific arguments to the wider study of the methodological rules governing science.
83. Ibid., pp. 378–382.
84. See *UQ*, pp. 84–85; see also *BGE*, pp. XIII and 441.
85. By contrast, a few years later Popper declared that "it took me a few years to notice that the two problems—of demarcation and of induction—were in a sense one" (*CR*, p. 52).

NOTES TO CHAPTER 2

1. *LSD*, p. 28; see also *CR*, pp. 42–59; and *RMC*, pp. 1013–1030.
2. *LSD*, p. 34; later he states it as "the problem of finding the criterion of the empirical character of science" (*ibidem*, p. 55n3). See also pp. 313–314, as well as *CR*, pp. 33–39, 42 and 253–292, and *RMC*, pp. 965–974 and 976–987.
3. See *BGE*, p. 4; and *LSD*, p. 34; see also *UQ*, pp. 78–87.
4. *OK*, p. 4.
5. *LSD*, p. 27: "no matter how many instances of white swans we may have observed, this does not justify the conclusion that *all* swans are white."
6. *OK*, p. 4.
7. Hume did not question the certainty of our theories: he firmly believed that we can be certain of our theories notwithstanding the method we used to achieve them (induction) is invalid. Such certainty, however, does not rely on reason, but rather on habit.
8. *RMC*, p. 1015. See also Miller (1994), chap. 2; and (2006), chap. 5.
9. *RMC*, p. 1015. As Miller (2006), pp. 10–11, observes: "it is not so much inductivism that [Popper] demolished, as justificationism: the view that the rationality of science (and other intellectual activities) consists in the pursuit

and attainment of (wholly or partially) justified truth, rather than of truth. [. . .] He encourages us to see science as a reckless human activity, whose extraordinary purchase on truth about this strange and beautiful world is not less genuine for being quite unsecured."

10. *OK*, p. 7.
11. Ibid.
12. Ibid., p. 9.
13. Ibid., p. 7.
14. Ibid.
15. Ibid., p. 8.
16. Ibid.
17. *LSD*, p. 31. Popper always insisted—and rightly so, of course—that the origin of ideas has nothing to do with their validity: the "context of discovery" and the "context of justification"—a distinction that can be traced back at least to John Herschel's *A Preliminary Discourse on the Study of Natural Philosophy* (1830), and was later resumed by Hans Reichenbach in *Experience and Prediction: An Analysis of the Foundation of Science* (1938)—must be kept separate and studied differently. What counts is intersubjective criticism, not freedom from prejudices. However, as the story of Popper's early intellectual development told in the previous chapter shows, questions of genesis are not theoretically trivial, and are especially welcome when they invite and facilitate criticism.
18. *LSD*, p. 32; see also *UQ*, pp. 143–146.
19. *LSD*, p. 33.
20. Ibid.
21. See *UQ*, pp. 35–38.
22. *CR*, p. 34.
23. Ibid., pp. 34–35.
24. Ibid., p. 36; see also *UQ*, pp. 31–38; as well as *P1*, pp. 163–174.
25. *LSD*, p. 34.
26. See *RMC*, pp. 1105–1114.
27. *LSD*, p. 41; see also p. 265; as well as *P1*, pp. 181–189.
28. *LSD*, p. 40; see also p. 279; as well as *CR*, p. 256.
29. See also Miller (2006), chap. 4.
30. *LSD*, pp. 40–41. See also *ALPS*, pp. 16–22; and *CR*, p. 37: "*the criterion of the scientific status of a theory is its falsifiability, or refutability, or testability.*" For a detailed discussion, see *P1*, chap. 2; see also *OS2*, pp. 259–260.
31. See *P1*, chap. 3; see also *CR*, chaps. 8 and 11; as well as *RMC*, pp. 1066–1072. In the late 1950s—while working on *LSD*, the revised and augmented English edition of *LF*—Popper would considerably develop the ideas first presented some twenty years earlier. The long reflections on the relevance and role of metaphysics for the growth of scientific theories would become the three volumes of the important *Postscript* that, after circulating only within the inner circle of Popper's friends and pupils for many years (under the title *Postscript: After Twenty Years*), was eventually seen through the press in 1982–1983.
32. More precisely, by "basic statement" (or "basic proposition") Popper refers to statements "which can serve as a premise in an empirical falsification; in brief, a statement of a singular fact" (*LSD*, p. 43). They have the form of singular existential statements: they state that, in a given region of space and time an observable event is taking place.
33. *LSD*, p. 86. Popper further distinguishes between *falsifiability* and actual *falsification* of a theory: the former is a requisite that guarantees the empirical character of a theoretical system, whereas the latter is a procedure

governed by the rules of method. A theory may be called scientific if it is falsifiable, that is, if it can be refuted by experience; it is falsified when it is agreed to accept the basic statements that contradict it. See *LSD*, pp. 86–87.

34. Ibid., p. 112.

35. Ibid., p. 113; a theory with a high degree of testability would be easier to falsify, "since it allows the empirical world only a narrow range of possibilities; for it rules out almost all conceivable, *i.e.* logically possible, events. It asserts so much about the world of experience, its empirical content is so great, that there is, as it were, little chance for it to escape falsification. Now theoretical science aims, precisely, at obtaining theories which are easily falsifiable in this sense. It aims at restricting the range of permitted events to a minimum; and, if this can be done at all, to such a degree that any further restriction would lead to an actual empirical falsification of the theory" (ibid.).

36. As Popper himself declares in the first pages of *LSD*, his criterion is to be regarded "as a *proposal for an agreement or convention*. As to the suitability of any such convention opinions may differ; and a reasonable discussion of these questions is only possible between parties having some purpose in common. The choice of this purpose must, of course, be ultimately a matter of decision, going beyond rational argument" (*LSD*, p. 37). Popper firmly believed that any two parties, merely sharing the concern for truth and willing to listen to each other's arguments, may fruitfully talk and progress (see *OS2*, pp. 224–247). His commitment to truth as a Kantian regulative idea bears an ethical dimension: see chap. 5.

37. Quine (1951), p. 37.

38. See Wittgenstein (1921), 4.063, 6.53, 6.54. Popper notices: "nothing is easier than to unmask a problem as 'meaningless' or 'pseudo': All you have to do is to fix upon a conveniently narrow meaning for 'meaning,' and you will soon be bound to say of any inconvenient question that you are unable to detect any meaning in it. Moreover, if you admit as meaningful none except problems in natural science, any debate about the concept of 'meaning' will also turn out to be meaningless. The dogma of meaning, once enthroned, is elevated forever above the battle: It can no longer be attacked. It has become (in Wittgenstein's own words) 'unassailable and definitive'" (*LSD*, p. 51; the reference is to Wittgenstein 1921, p. 5).

39. Wittgenstein (1921), 4.112.

40. Ibid., 4.113.

41. Ibid., 4.114.

42. Ibid., 4.115.

43. See Waismann (1930), p. 5: "If there is no way of telling when a proposition is true, then the proposition has no sense whatever; for the sense of a proposition is the method of its verification"; the expression might have originated from a conversation Waismann had with Wittgenstein: see Waismann (1967), pp. 47–48; and (2003), pp. 116–121 and 492–497. See also Schlick (1936), p. 458: "Stating the meaning of a sentence amounts to stating the rules according to which the sentence is to be used, and this is the same as stating the way in which it can be verified (or falsified). The meaning of a proposition is the method of its verification."

44. See *CR*, pp. 39–40.

45. See Carnap (1932a).

46. *CR*, p. 40.

47. *CR*, p. 50; see also pp. 253 and 281: "against the intention of its defenders, *it did not exclude obvious metaphysical statements; but it did exclude the*

most important and interesting of all scientific statements, that is to say, the scientific theories, *the universal laws* of nature."

48. *LSD*, p. 36.
49. Ibid., p. 37.
50. See Agassi (1964a), as well as his (1968) and (1975).
51. For other examples, see *P1*, pp. 191–193; and *P3*, pp. 162–165.
52. *LSD*, pp. 277–278; see also *CR*, chap. 8.
53. *P1*, p. 192.
54. See *P1*, pp. 189–193; *P3*, pp. 159–177; as well as *UQ*, pp. 148–151; see also Agassi (1964a), and next chapter.
55. See *CR*, p. 199.
56. *CR*, p. 199; see the whole discussion on pp. 197–200.
57. See Watkins (1958), (1975), and (1978).
58. Agassi (1964b), p. 272. "In my view," he continues, "the interaction between physics and metaphysics is by way of metaphysics prescribing programmes for future scientific development" (ibid.).
59. See Kuhn (1962), p. 184.
60. See Lakatos (1970), pp. 132–133. On the relationship between Lakatos' "scientific research programmes" and Popper's (as well as Agassi's) "metaphysical research programmes" see Bartley (1976); Berkson (1976); and Wettersten (1992), pp. 241–243.
61. *LSD*, p. 41.
62. *LSD*, p. 41.
63. Ibid. Later on, Popper writes: "a theory makes assertions only about its potential falsifiers. (It asserts their falsity). About the 'permitted' basic statements it says nothing" (p. 86); and again: "the amount of empirical information conveyed by a theory, or its *empirical content*, increases with its degree of falsifiability" (p. 113; see also pp. 112–113). On the distinction between empirical content and logical content of a theory, see pp. 119–121.
64. Ibid., p. 79.
65. Observation, Popper agrees, is theory-laden: "theory dominates the experimental work from its initial planning up to the finishing touches in the laboratory" (*LSD*, p. 107; this passage was already present in *LF*, p. 63). In a footnote added to the English edition, he continues: "observations, and even more so observation statements and statements of experimental results, are always *interpretations* of the facts observed; [. . .] they are *interpretations in the light of theories*" (ibid., p. 107*3: this is the reason why, Popper continues, it is always "deceptively easy" to find verifications of a theory). And again: "our ordinary language is full of theories; [. . .] observation is always *observation in the light of theories*; [. . .] it is only the inductivist prejudice which leads people to think that there could be a phenomenal language, free of theories, and distinguishable from a 'theoretical language'" (p. 59n1). See also ibid., pp. 42, 50n1, 81–87, 106–107, 280, 412–413, and 423; as well as *OS2*, 213–214 and 260–261; *CR*, pp. 38n3, 44–52, and 387 (where Popper speaks of facts as "soaked in theory"); and *ALPS*, pp. 6 and 33–34.
66. *LSD*, p. 80.
67. Ibid., pp. 80–81.
68. See *LSD*, pp. 41–42 and 50; as well as *P1*, pp. xxi–xxiii.
69. As we have seen in the previous chapter, the challenge of conventionalism played a key role in this passage—a passage that can be traced back to the shift from Popper's considerations in *BGE* (without §11) to those in §11 and *LF*.
70. *LSD*, p. 49.
71. Ibid.

72. Ibid., p. 82.
73. *LSD*, p. 49; see also p. 98n1: "How can we best *criticize* our theories (our hypotheses, our guesses) rather than defend them against doubt?"
74. Ibid., p. 42.
75. Ibid., p. 50; and again: "*all rational criticism is criticism of the claim of a theory to be true, and to be able to solve the problems which it was designed to solve.* Thus I do not *replace* the question whether a theory is true by the question whether it is better than another. Rather, I replace the question whether we can produce valid *reasons* (positive reasons) in favour of the truth of a theory by the question whether we can produce valid *reasons* (critical reasons) against its being true, or against the truth of its competitors" (*P1*, pp. 24–25).
76. *LSD*, p. 53.
77. Ibid.; see also p. 280. This passage may very well translate into political terms, thus showing how the two spheres of Popper's philosophical reflection—the scientific enterprise, on the one hand, politics and society, on the other—are tightly interwoven. Just as in science there is no absolutely true theory so, in politics, there is no perfect society; just as science may advance by appealing to ever better but never ultimate theories, so society may evolve by adopting ever better forms and political assets, but these are never final. Just as in science, so in society growth is neither necessary or predictable, but depends on the effectiveness of institutions designed by rulers. Just, as in science, the problem is not the elimination but the correction of errors, without appealing to ultimate foundations or undisputable authorities, so, in the political arena, the central problem becomes that of establishing institutional checks for political choices. This view would imply a rigorous distinction of political powers (the checking role of parliament, on the one hand, and the executive role of government, on the other), the temporariness of any elective office and the appeal to every citizen's own involvement and personal responsibility: for—to use Pericles' words Popper took as the motto of *OS1*—"Although only a few may originate a policy, we are all able to judge it" (see also *OS1*, pp. 185–189).
78. *LSD*, p. 54.
79. Ibid., p. 56.
80. "Objectivity is not the result of disinterested and unprejudiced observation. Objectivity, and also unbiased observation, are the result of criticism, including the criticism of observational reports. For we cannot avoid or suppress our theories, or prevent them from influencing our observations; yet we can try to recognize them as hypotheses, and to formulate them explicitly, so that they may be criticized" (*P1*, p. 48).
81. *LSD*, p. 44; Popper then generalizes this formulation in n1: "inter-subjective *testing* is merely a very important aspect of the more general idea of inter-subjective *criticism*, or in other words, of the idea of mutual rational control by critical discussion"; see also pp. 81–84 and 97–100. The idea is further discussed in *OS2*, chaps. 23–24; and *PH*, pp. 155–159.
82. *LSD*, pp. 82–83.
83. Duhem's idea was that "An experiment in physics can never condemn an isolated hypothesis but a whole theoretical group" (Duhem 1906, p. 183). See also pp. 183–190, 199–200, 208, 216, 220, 258, 278, and particularly pp. 187–188: "when the experiment is in disagreement with [the physicist's] predictions, what he learns is that at least one of the hypotheses constituting this group is unacceptable and ought to be modified; but the experiment does not designate which one should be changed. [. . . .] People generally think that each one of the hypotheses employed in physics can be taken in isolation, checked

by experiment, and then, when many varied tests have established its validity, given a definitive place in the system of physics. In reality, this is not the case. Physics is not a machine which lets itself be taken apart; we cannot try each piece in isolation and, in order to adjust it, wait until its solidity has been carefully checked. Physical science is a system that must be taken as a whole; it is an organism in which one part cannot be made to function except when the parts that are most remote from it are called into play, some more so than others, but all to some degree." Duhem took up a consideration of Poincaré, see his (1902), pp. 151–152: "If we construct a theory based upon multiple hypotheses, and if experiments condemn it, which of the premises must be changed? It is impossible to tell." Years later, in a different context (the discussion of reductionism, one of the "two dogmas of empiricism"), Quine developed and specified Duhem's methodological holism, advancing a new version of the thesis: "our statements about the external world face the tribunal of sense experience not individually but only as a corporate body" (Quine 1951, p. 41). See also pp. 40–41. Accordingly, the thesis is commonly referred to as the "Duhem-Quine thesis": *pace* Lakatos, see his (1970), pp. 180–189, Popper knew this issue very well and considered it in *LSD*, pp. 76–77. See also Harding (ed.) (1976), and Gillies (1993), chap. 10. Neurath held a similar view: see, for example, his (1933), p. 92 (Neurath's recurrent simile first appeared in print in 1913).

84. *LSD*, p. 83. See also *OS2*, pp. 260–261: "Indeed, the theory or hypothesis could be described as the crystallization of a point of view. For if we attempt to formulate our point of view, then this formulation will, as a rule, be what one sometimes calls a working hypothesis; that is to say, a provisional assumption whose function is to help us to select, and to order, the facts. But we should be clear that there cannot be any theory or hypothesis which is not, in this sense, a working hypothesis, and does not remain one. For no theory is final, and every theory helps us to select and order facts." Other responses to Duhem's thesis are to be found in *LSD*, pp. 71–72 and 75–77; *BGE*, pp. 257–263; and *P1*, pp. 181–189. See also Miller (2006), pp. 86, 97, and 108–109.

85. See *LSD*, pp. 93–95.

86. The term was coined by Neurath in his (1932), p. 393. See also Carnap (1932b) and (1933a).

87. See Carnap (1932b), pp. 42–52. See also Neurath (1933) and Carnap (1933b). Unlike Carnap, Neurath thought that protocol statements (or sentences) may be revised. Popper took it as a step in the right direction, "but it leads nowhere if it is not followed up by another step: we need a set of rules to limit the arbitrariness of 'deleting' (or else 'accepting') a protocol sentence. Neurath fails to give any such rules and thus unwittingly throws empiricism overboard. For without such rules, empirical statements are no longer distinguished from any other sort of statements" (*LSD*, p. 97; see also the discussion on pp. 96–97).

88. See *LSD*, pp. 104–111.

89. "Experiences can *motivate a decision*, and hence an acceptance or rejection of a statement, but a basic statement cannot be *justified* by them—no more than by thumping on the table" (*LSD*, p. 105).

90. *LSD*, p. 108; on instrumentalism, see p. 59n*1, as well as p. 61, addition * to n1; and *P1*, pp. 111–131.

91. *LSD*, p. 109. Moreover, "basic statements are not justifiable by our immediate experiences, but are, from the logical point of view, accepted by an act, by a free decision" (ibid.).

92. On simplicity, see *LSD*, chap. 7, and particularly pp. 144–145, where the connection between simplicity and severity of tests is discussed.

93. Ibid., p. 109.
94. Ibid., p. 111.
95. Ibid., pp. 251–265.
96. Ibid., p. 251. "Corroboration" and "degree of corroboration," it is to be noted, do not equal to "confirmation" and "degree of confirmation" (that is, probability) in the sense of Carnap (1936–1937).
97. See *CR*, pp. 240–248; see also the detailed discussion in *P1*, pp. 233–244; as well as *UQ*, pp. 98–104.
98. *LSD*, p. 266.
99. Ibid., p. 267. And "the severity of the tests, in its turn, depends upon the degree of testability, and thus upon the simplicity of the hypothesis: the hypothesis which is falsifiable in a higher degree, or the simpler hypothesis, is also the one which is corroborable in a higher degree" (ibid.). See also Appendix *9.
100. *LSD*, p. 270.
101. See *CR*, pp. 256–257.
102. *LSD*, p. 270n3; see also *P1*, pp. 236–243.
103. "*The logical probability of a statement is complementary to its degree of falsifiability*: it increases with decreasing degree of falsifiability. The logical probability 1 corresponds to the degree 0 of falsifiability, and *vice versa*" (*LSD*, p. 119). See also *CR*, pp. 285–287.
104. *CR*, p. 286. The statement "tomorrow it will rain or it will not rain" is certainly true, and therefore its logical probability is 1. There is no evidence whatsoever that might clash with it, or contradict it, and therefore its degree testability is 0. However certain, though, it is not informative: it does not tell us, for instance, whether we would better take the umbrella with us tomorrow.
105. See *P1*, pp. 232–233.
106. *UQ*, p. 102. See also *SIB*, pp. 9–10.
107. *UQ*, pp. 103–104.
108. As we have seen, Popper's criterion of demarcation does not distinguish between refuted and corroborated hypotheses. By the criterion of falsifiability even hypotheses which were amply refuted by experience are fully entitled to the honorary status of scientific hypotheses.
109. See *CR*, chap. 10, especially 240–248. A detailed treatment of this issue is to be found in *P1*, part 1, chap. 4.
110. *LSD*, p. 215n.
111. *CR*, p. 241.
112. Ibid., p. 242.
113. Ibid., p. 247. See also Watkins (1984), pp. 288–304.
114. See Agassi (1968), pp. 26–27.
115. *RMC*, p. 1193.
116. Agassi (1961).
117. *CR*, p. 248n31.
118. Newton-Smith (1981), p. 68.
119. Miller (1994), pp. 120–121.
120. Agassi (1968), p. 27; see also Agassi (1993), pp. 201–204.
121. Miller (1994), p. 120. "We might put it this way. When a theory fails a test, we learn something but end up knowing nothing (since what we knew, our theory, has been eliminated). But when a theory passes a test (when, that is to say, it is corroborated), we learn nothing (since we already knew what the result of the test was going to be) but we continue to know something" (ibid.).
122. *LSD*, p. 280.

123. Ibid.
124. *OS2*, p. 260. See also p. 261: "our description will always be incomplete, a mere selection, and a small one at that, of the facts which present themselves for description. This shows that it is not only impossible to avoid a selective point of view, but also wholly undesirable to attempt to do so; for if we could do so, we should get not a more 'objective' description, but only a mere heap of entirely unconnected statements."
125. See *OK*, pp. 60–65 and 341–361; as well as *RMC*, pp. 1016–1017.
126. Novalis (1960), p. 668. It is interesting to read also the rest of the passage Popper is quoting from, which continues: "Hasn't America been discovered with a hypothesis? / Long live hypothesis—only she remains / eternally new, though it often defeats itself" (ibid.).
127. *LSD*, p. 59. See also *P2*, pp. 41–43.
128. See *CR*, pp. 223–228; *OK*, pp. 44–52; *UQ*, pp. 98–100 and 141–143; as well as *P1*, p. 26. On critical rationalism's debt to Tarski, see Miller (2006), chap. 9.
129. *LSD*, p. 278.
130. *LSD*, p. 280. Popper continues: "It may indeed be corroborated, but every corroboration is relative to other statements which, again are tentative" (ibid.).
131. "The natural result of any investigation is that the investigators either discover the object of search, or deny that it is discoverable and confess it to be inapprehensible, or persist in their search. So, too, with regard to the objects investigated by philosophy, this is probably why some have claimed to have discovered the truth, others have asserted that it cannot be apprehended, while others again go on inquiring. [. . .] The skeptics keep on searching" (Sextus Empiricus [1933], book 1, chap. 1, pp. 1–3). See also Miller (2006), chap. 7.
132. "The central problem of epistemology has always been and still is the problem of the growth of knowledge" (*LSD*, p. 15).
133. Another term is "approach (or closeness, or similarity) to the truth." See *CR*, pp. 228–237; *OK*, pp. 52–60; and *RMC*, pp. 1100–1103.
134. The notions of objective truth and approach (or getting nearer) to the truth are "of great help in analysing the growth of knowledge" (*CR*, p. 216).
135. See *CR*, pp. 233–234; see also *OK*, pp. 47–52; and *MF*, pp. 174–176.
136. See Miller (1974a), (1974b), and (1976); Tichý (1974), (1976), and (1978); and Harris (1974). See also Miller (1994), chap. 10; as well as Niiniluoto (1987).
137. See *P1*, pp. xxxv–xxxvii; and *OK*, pp. 367–372.
138. *OK*, p. 59, see also pp. 314–340; as well as *RMC*, pp. 1091–1095 and 1103–1105.
139. *CR*, p. 217.
140. Ibid.; and *OK*, pp. 54–58.
141. *CR*, p. 218. Closeness to the truth (that is, verisimilitude) and closeness to certainty (that is, probability) are to be sharply separated: see ibid., pp. 233–237.
142. Ibid., p. 218.
143. Ibid., pp. 217–220
144. *LSD*, p. 146.
145. Ibid., pp. 146–147; as well as *UQ*, pp. 99–103.
146. *LSD*, p. 148.
147. Ibid.; see also *CR*, pp. 280–292. If a follows from b, $p(a|b) = 1$; if a contradicts b, $p(a|b) = 0$; in all other instances, $p(a|b)$ is higher the less its content exceeds the content of b, on which the probability of a depends. Therefore the probability relation is a special kind of logical relation between two assertions. See *LSD*, pp. 148–150.

148. *LSD*, p. 149. "According to this interpretation, the statement 'The probability of the next throw with this die being a five equals 1/6' is not really an assertion about the next throw; rather, it is an assertion about a whole *class of throws* of which the next throw is merely an element. The statement in question says no more than that the relative frequency of fives, within this class of throws, equals 1/6" (ibid.). Later Popper would regard the statistical, or frequency interpretation as just the simplest among subjective interpretations: see Popper (1959), p. 26. And in Appendix *6 to *LSD* he would find the "measure-theoretical approach" preferable to the frequency interpretation: see *LSD*, p. 361. See also *P1*, part 2, chap. 3.
149. *LSD*, p. 190; see pp. 190–191.
150. Ibid., p. 190.
151. Ibid., p. 191.
152. Ibid., my emphasis.
153. Ibid., pp. 209–212.
154. See Popper (1957b), p. 66.
155. *LSD*, p. 210. Popper defines a probability statement as "formally singular" when it ascribes a probability to a single occurrence" (ibid., p. 209).
156. *P1*, p. 387.
157. See Popper (1957b), pp. 66–67.
158. See Popper (1959), pp. 31–35; and *P1*, pp. 352–355.
159. Popper (1959), p. 34.
160. Ibid., p. 35. See also *UQ*, pp. 151–155; as well as *RMC*, pp. 1117–1120 and 1125–1144. A most detailed treatment is to be found in *P1*, part 2.
161. *P1*, p. 287; see also pp. 358–359.
162. Ibid., p. 287.
163. *P1*, p. 347.
164. Ibid., p. 351.
165. Ibid., pp. 351–352.
166. Ibid., p. 360; see also *P3*, pp. 151–156.
167. See *P3*, pp. 35–41; the whole discussion is on pp. 35–95.
168. Popper (1957b), p. 65; see also *P3*, pp. 205–211, and the "thirteen theses" expounded on pp. 46–86.
169. *P1*, p. 361.
170. Ibid.
171. *WP*, p. 12.
172. Ibid., p. 14. Such "situational" features are crucial for a realistic interpretation of the quantum theory.
173. Ibid., p. 17.
174. Ibid.
175. Ibid., p. 18.
176. Ibid.
177. Ibid., pp. 18–19.

NOTES TO CHAPTER 3

1. See, for instance, *ALPS*, chap. 7.
2. The term, of course, is to be understood in the meaning Popper gave it with his criterion of demarcation: as we have seen in the previous chapter, a doctrine is metaphysical if it is not falsifiable, or testable, and therefore cannot be refuted by appealing to experience (see *LSD*, pp. 34–42; or *P1*, part 1, chap. 3).

3. *LSD*, p. 15. And he goes on to say: "For me, at any rate, both philosophy and science would lose their attraction if they were to give up that pursuit" (ibid.). See also *P3*, p. 1.

4. *P3*, p. 161. Research programmes such as Parmenides' block universe, atomism, the clockwork theory of the world, field theories, or the statistical interpretation of quantum physics, are "indispensable for science" (*P3*, p. 165). "Originally they were all metaphysical in nearly every sense of the word (although some of them became scientific in time); they were vast generalizations, based upon various intuitive ideas, most of which now strike us as mistaken. They were unifying pictures of the world—the real world. They were highly speculative; and they were, originally, non-testable. Indeed they may all be said to have been more of the nature of myths, or of dreams, than of science. But they helped to give science its problems, its purposes, and its inspiration" (ibid.).

5. *P3*, p. 161.

6. Ibid. See also *P1*, pp. 131–149.

7. *P3*, p. 210.

8. Ibid., p. 211.

9. Ibid.

10. See *RMC*, p. 963; as well as *OK*, pp. 35–44 and chap. 8; and *UQ*, pp. 90–94.

11. *OK*, p. 35.

12. Ibid., p. 38.

13. See *OK*, pp. 38 and 105; as well as *P1*, pp. 82–83.

14. *OK*, p. 39.

15. Ibid., p. 40. See also *P1*, pp. 83–84.

16. "We accept things as 'real' if they can causally act upon, or interact with, ordinary real material things" (*SIB*, p. 10).

17. *OK*, p. 41.

18. Ibid.

19. Ibid.

20. Ibid., pp. 41–42.

21. Ibid., p. 42.

22. Ibid. See also *LSD*, pp. 280–281; as well as *OS2*, pp. 224–225; and *CR*, p. 334. Science, Popper never tires to repeat, is the quest for truth, not for certainty: see *ALPS*, p. 38; and *SBW*, pp. 4 and 38.

23. *P3*, pp. 198–199. See also *P2*, p. 8; *RMC*, pp. 1053–1059; *UQ*, pp. 94–95; and *SIB*, pp. 32–35.

24. To the idea of an "open universe" Popper devotes the whole of *P2*, and the idea is present in the titles of some of his later works as well, not to mention Popper (1950) and *OK*, especially chap. 6). The defence of creativity and human freedom constituted the key issue of both *PH* and *OS*. John Watkins, Popper's pupil and successor at the London School of Economics and Political Science, picks indeterminism as one of the two fundamental features of Popper's philosophy: together with the critical attitude, it allows us to view Popper's multifaceted philosophical production as a single, coherent whole. See Watkins (1974), p. 372: "I take as the original core of Popper's philosophy his *falsificationism*; and I take his *indeterminism* to have been the most striking component of his metaphysical outlook around 1950. [. . .] I shall argue that his indeterminism is significantly related to his evolutionism, which in turn is significantly related to his falsificationism."

25. *P2*, p. 5.

26. *P2*, p. xx. See also pp. 1–2, where determinism is defined as "the doctrine that the structure of the world is such that *any event can be rationally pre-*

dicted, with any desired degree of precision, if we are given a sufficiently precise description of past events, together with all the laws of nature."

27. *P2*, p. xx.

28. *P2*, p. 5. Although this kind of determinism initiates in Greek classical thought, it found a rigorous definition only in Pierre Simon de Laplace's *Philosophical Essay on Probability* (1814): since the world consists of particles acting upon one another according to the laws of Newtonian mechanics, a complete and precise knowledge of the state of the world system at any given moment would allow to deduce its state in any other (future) moment. However, since such a knowledge would exceed human possibilities, Laplace appealed to a fictitious demon, which no longer represented almighty God, but only a sort of super-human intelligence, or super-scientist. In so doing, Laplace *"makes the doctrine of determinism a truth of science rather than a religion"* (*P2*, p. 30).

29. *P2*, p. 6. Popper offers scientific determinism also as a criticism of the commonsense idea according to which all events can be divided into two kinds: those that can be predicted, such as the cycle of seasons, the motions of the planets, and the functioning of a clock; and those that cannot be predicted, such as the vagaries of the weather or the shape and behaviour of clouds: see ibid., pp. 6–7 and 18–19; as well as *OK*, pp. 207–234.

30. *P2*, p. 8.

31. Ibid., p. 9.

32. Ibid., p. 10.

33. Ibid., pp. 27–28 and 41.

34. Ibid., pp. 55–56.

35. Ibid., p. 58.

36. Ibid., p. 59.

37. Ibid., p. 61. "It is not the kicks from the back, from the past, that *impel* us but the attraction, the lure of the future and its competing possibilities, that *attract* us, that *entice* us. That is what keeps life—and, indeed, the world—unfolding" (*WP*, pp. 20–21).

38. *P2*, p. 62. From this very consideration Popper moves to criticize historicism: see *P2*, pp. 62–64; as well as *PH*, pp. v–vii.

39. See *P2*, p. 65.

40. Ibid., pp. 89–92; as well as *UQ*, pp. 128–131; see also *TWP*, chaps. 3–6.

41. See *P2*, pp. 90–91.

42. Ibid., p. 91.

43. Ibid., pp. 91–92.

44. *OK*, p. 215; see also *SIB*, pp. 33–34; as well as the detailed discussion in *OK*, chap. 6.

45. See *OK*, p. 230; as well as *P2*, pp. 113–130.

46. *P2*, p. 114; see also p. 130.

47. Popper briefly refers to the existence of three worlds, or universes, already in a lecture delivered in 1953. Taking the cue from Karl Bühler's *Sprachtheorie* (1934), he argues for "the impossibility of a physicalistic causal theory of the human language" (*CR*, p. 293) and defends a dualist stance as to the body-mind relationship. Whereas Descartes postulated the existence of two substances—*res cogitans* and *res extensa*—Popper argues for the existence of two states, the mental and the physical, mutually interacting (see *OK*, p. 231n43). At any rate, he attributes such states to bodies and minds and constructs universes containing them (see *CR*, chap. 12). To these he adds objects of another kind: "Logical relationships, such as consistency, do not belong to the physical world"; and since they do not belong to the subjective thought either, "They are abstractions (perhaps 'products of the mind')"

(*CR*, p. 298). However, it is in his 1966 Arthur Holly Compton Memorial Lecture that Popper explicitly postulates a *"universe of abstract meanings"* (*OK*, p. 230).

48. *OK*, p. 106. This is a frequently discussed topic in Popper's later productions. Apart from the references given here, see *OK*, pp. 153–165; *SIB*, chap. P2 and pp. 537–547; *RMC*, 1050–1053 and 1072–1078; *UQ*, pp. 180–187; *ALPS*, pp. 24–35; *SBW*, pp. 7–9, 21–26, and 161–166.

49. In positing these "worlds," Popper is not advancing any scientific claim: World 1, World 2, and World 3 do not belong to natural science, but rather to metaphysics. It is, however, a crucial issue for Popper's critical rationalism: by declaring theories to be not merely a psychic product of individuals, but something that can be expressed and shared by others, Popper allows for the possibility of criticism.

50. See *OK*, pp. 158–161; *SBW*, pp. 7–10; *SIB*, pp. 38–47 and 548–552. "Of course, the name 'world 3' is a metaphor: we could, if we wish to, distinguish more than three worlds. We could, for example, distinguish the world of objective knowledge as a separate world from that of the arts, and other distinctions would also be possible" (*KBMP*, p. 25; see also pp. 118–119).

51. See *OK*, pp. 155–156.

52. See *OK*, pp. 122–127 and 153–154; as well as *SBW*, pp. 161–165.

53. Popper himself acknowledges Plato the merit to have called the philosophers' attention to "an objective, autonomous third world of logical contents" (*SBW*, pp. 161–162).

54. See *OK*, pp. 112, 115–119, and 159–160; *KBMP*, pp. 30–31 and 118–119; and *SBW*, pp. 163–164.

55. See *OK*, pp. 122–123 and 300–301; and *SIB*, pp. 38–39.

56. *KBMP*, p. 49; see also p. 50.

57. *OK*, pp. 125–126.

58. Ibid., p. 126.

59. See *OK*, pp. 126–127.

60. See *KBMP*, pp. 50–51

61. Already in 1902, Frege—analogously to Popper—distinguished the psychological from the logical aspects of thought, referring the former to the subjective processes of thought, whilst referring the latter to its objective contents. However, such considerations were made explicit only in Frege (1919), when he introduced the idea of "the third realm," the realm of thought in the objective sense.

62. See *OK*, pp. 157–158; as well as *SBW*, pp. 161–162.

63. See *UQ*, p. 188; *OK*, p. 127; *ALPS*, pp. 8–9 and 25; and *KBMP*, p. 51.

64. *OK*, pp. 108–109; see also pp. 212–214, 64–67, 72–78, and 156–158; as well as *KBMP*, pp. 3–4; and *ALPS*, p. 8.

65. At first, Popper used expressions such as "first world," "second world," and "third world" instead of the later "World 1," "World 2," and "World 3" (the latter terminology being suggested by John C. Eccles).

66. *OK*, p. 111.

67. Ibid. The world of objective knowledge, Popper concludes, "is of decisive importance for epistemology" (ibid.).

68. See also the most interesting remarks in Bartley (1990), pp. 31–38.

69. *OK*, p. 112.

70. Ibid.

71. Ibid., p. 114.

72. Ibid.

73. *KBMP*, p. 24.

74. Ibid., p. 4.

75. *OK*, p. 73.
76. Ibid., p. 109.
77. See *SIB*, pp. 14–21.
78. Ibid., p. 99.
79. *KBMP*, p. ix. See also *UQ*, pp. 187–193; *ALPS*, chap. 2; and *P2*, pp. 153–159.
80. *KBMP*, p. 7.
81. See *KBMP*, pp. 4–5; as well as *UQ*, pp. 188–193.
82. See *P2*, 155–156; and *ALPS*, pp. 29–31. For a detailed history of the problem, see *SIB*, chap. P5.
83. *P2*, p. 155.
84. Ibid.
85. *OK*, p. 231. The second half of the problem is deemed less important, for Popper is mainly concerned with the explanation of the influence of World 2 upon World 1.
86. Ibid., p. 230.
87. Ibid., p. 231.
88. Ibid., pp. 231–232.
89. Ibid., p. 232; see also *KBMP*, pp. 111–112.
90. *ALPS*, p. 24.
91. See *P2*, pp. 131–162. Popper calls himself a "trialist."
92. *OK*, p. 252. Popper acknowledges he has not solved the problem of how such interaction takes place, and indeed thinks it can perhaps never be solved. See *OK*, p. 255; as well as *ALPS*, p. 23: "it is the deepest and most difficult problem of philosophy, the central problem of modern metaphysics" (see pp. 23–24).
93. *KBMP*, pp. 114–115.
94. See *SIB*, pp. 22–35 and 437–448.
95. See *KBMP*, p. 115; as well as *CR*, chaps. 12–13.
96. *OK*, p. 119, see also pp. 164–165; as well as *KBMP*, pp. 10–13, 61–64, and 79; *UQ*, pp. 132–135; *ALPS*, pp. 3–15; and *MF*, p. 101 and pp. 140–141. The tetradic schema might be elaborated in various ways, for example by considering various and competing tentative theories (TT_1, TT_2, . . . TT_n) advanced in the attempt to solve the initial problem, each of which gives rise to new tests and goes through a different stage of critical discussion (EE_1, EE_2, . . . EE_n), thus leading to different new problems (P_1, P_2, . . . P_n). The whole process might eventually merge in a "critical evaluative discussion," where attempts are made to decide which of the competing theories bear the clash with tests and survive, and which is to be eliminated. Such evaluation "is always *critical*, and its aim is the discovery and *elimination of error*" (*OK*, p. 144).
97. *KBMP*, p. 11: "almost every solution opens up in its turn a whole new world of open problems" (*P2*, p. 162; see also p. 109).
98. *OK*, p. 144.
99. See *SIB*, pp. 451–452.
100. As a consequence, the "objectivity" that characterizes World 3 objects and the knowledge that derives from it is, more precisely, an intersubjectivity. Lacking the access to reality as it really is (Kant's *Ding an sich*) man cannot help but formulate hypotheses and improve them with the help of others.
101. See *OK*, 119–121; *CR*, pp. 134–135 and 295; *KBMP*, pp. 83–88; *SIB*, pp. 57–60 and 455–456; *P2*, pp. 82–83 and 153–154; *SBW*, pp. 21–22; and *ALPS*, pp. 40–41.
102. See *OK*, p. 121; and *CR*, pp. 64–65.
103. *OK*, p. 121.

104. Ibid.
105. *KBMP*, p. 13.
106. Ibid., p. 79. "All life is problem solving. All organisms are inventors and technicians, good or not so good, successful or not so successful, in solving technical problems" (*ALPS*, p. 100).
107. Once again, Popper appeals to Novalis' words, as he did at the opening of *LF*: "Hypotheses are nets: only he who casts will catch."
108. *OK*, pp. 145–146.
109. Ibid., p. 147.
110. Ibid., p. 148.
111. *KBMP*, p. 12. See also *OK*, pp. 67–71 and 256–272; *SIB*, pp. 11–14, 48–49, and 120–135; *UQ*, pp. 167–180; *RMC*, 1059–1065; *ALPS*, pp. 5–7, 38–40, and 45–54; and *MF*, pp. 2–11.
112. *KBMP*, p. 12; see also *OK*, pp. 148–149; and *MF*, pp. 7 and 69.
113. See *SIB*, p. 558; as well as the essay comprised in two complementary collections with a significant title: *ALPS*, particularly pp. 38–39 and 100–102; and *SBW*, particularly pp. vii–viii.
114. *OK*, p. 148.
115. Ibid., p. 149.

NOTES TO CHAPTER 4

1. Popper (1970), p. 51. In some notes taken for a possible revised edition of *Criticism and the Growth of Knowledge*, Popper softens his remark, describing Kuhn's criticism as "one of the most interesting." Popper replies to Kuhn's criticism in his (1970); (1975); *RMC*, pp. 1144–1153; *P1*, pp. xxxi–xxxv; and *MF*, pp. 33–64. There are a few interesting notes collected in Popper Archive (75.5–10), as well as an unfinished essay especially devoted to Kuhn, "Revolution and Continuity in Science," dated February 6, 1972, in Popper Archive (120.11); most likely, the latter was a draft of Popper's reply to Kuhn that eventually appeared in *RMC*, pp. 1144–1148.
2. Friedman (1991), p. 1.
3. See Carnap's letters to Kuhn, published in Reisch (1991); for the relationship between Carnap and Kuhn, see also Irzik and Grünberg (1995); Friedman (1991); (1992); (1993); and (2001), pp. 18–19, 22, 41–43, and 56; Earman (1993); Axtell (1993); Irzik (2002); and Gattei (forthcom.), chap. 5. Indeed, in a recent article Michael Friedman explicitly outlines his project to rescue Logical Positivism through a reassessment of Kuhn's philosophy: see his (2003).
4. See Kuhn (1962), pp. 170–173. In fact, in the first edition of *The Structure of Scientific Revolutions* the term "truth" appears only in a quotation from Francis Bacon (on p. 18).
5. Ibid., pp. 170–171.
6. Ibid., p. 171. In his (2003) Brendan Larvor argues that in *The Structure of Scientific Revolutions* Kuhn allowed some features of his procedure and experience as a historian of science to pass over into his model for the growth of scientific knowledge: the fact that science is partly directed by extrascientific factors, incommensurability, the absence of any ahistorical standard of rationality and, most particularly, that science cannot be shown to be heading towards the truth—all these appear as methodological commitments rather than historical-philosophical theses. This reading of Larvor's fits very well with Feyerabend's 1960–1961 charge to Kuhn of deliberately blending descriptive and prescriptive elements in his book: see Feyerabend

(1995), especially pp. 355, 360, and 366–368; (2006), pp. 614–618; see also Feyerabend (1970a), pp. 198–199; and (1978), pp. 155–156.

7. Kuhn (1962, 1970), p. 206; see also pp. 205–207.

8. See Kuhn (1989), (1991), (1992), and (1993).

9. Kuhn (1962, 1970), p. 206. Such expressions like "zeroing in on the truth," or "getting closer and closer to the truth" are meaningless as a consequence of incommensurability. See also Kuhn (1993), pp. 243–244.

10. A continuous development lacks "a fixed, rigid Archimedean platform [that] could supply a base from which to measure the distance between current belief and true belief. In the absence of that platform, it's hard to imagine what such a measurement would be, what the phrase 'closer to the truth' can mean" (Kuhn 1992, p. 115).

11. The notion of necessary truth may be replaced by "something like a redundancy theory of truth" (Kuhn 1991, p. 99). See also ibid., pp. 95–99; (1992), p. 115; and (1993), pp. 244–245.

12. "Lexicon" is Kuhn's later term for "paradigm", thus stressing its linguistic features: a lexicon is a structured vocabulary comprising a taxonomy of kind terms that mirror and organize the objects in the world; for a detailed discussion, see Gattei (forthcom.), ch. 4.

13. See Wittgenstein (1969), §205: "If the true is what is grounded, then the round is not true, nor yet false"; and "why should the language-game rest on some kind of knowledge?" (§477; see also §559). According to Wittgenstein, a language game presents no gaps, since together with its possible moves it also defines the space which makes those very moves possible: just as the rules of the game define which moves belong to it, so the grammar of the language circumscribes what is meaningful. Nothing meaningful can therefore remain outside its boundaries and establish itself as a mark of the incompleteness of the language game (incommensurability). A game to which new rules are added is not a richer game, but simply a new game (paradigm shift). Therefore, a language game is criterion to itself—like the sample standard metre unit preserved at The International Bureau of Weights and Measures of Sèvres, near Paris, it is not itself measurable, since it is not possible to measure what is to be the unit of measurement: its having a length cannot be ascertained, but it is a feature which displays itself in the way we use it when measuring (see Wittgenstein (1953), part 1, §50, and §241). "If you tried to doubt everything you would not get as far as doubting anything. The game of doubting itself presupposes certainty" (Wittgenstein 1969, §115; see also §160, §450, and §625). Doubting the paradigm means, on the one hand, condemning oneself to silence; on the other, it means extending the practice of doubt beyond what is reasonable (i.e., meaningful) to doubt. Stephen Toulmin, a pupil of Wittgenstein's, thinks along the same line: "There must always be some point in a scientist's explanations where he comes to a stop: beyond this point, if he is pressed to explain further the fundamental basis of his explanation, he can say only that he has reached rock-bottom" (Toulmin 1961, p. 42).

14. Kuhn (1993), p. 244. Kuhn made this concept quite explicitly already in *The Structure of Scientific Revolutions*: "there is no standard higher than the assent of the relevant community" (Kuhn 1962, p. 94).

15. Putnam's "internal realism" closely resembles this stance: see, in particular, Putnam (1978), part 4; (1981), chaps. 5–7, and especially chap. 3; (1983), chaps. 2, 11, and 13; (1987); and (1990), part 1. In his introduction to the latter collection, James Conant remarks: "Having originally stood for the dream of realizing our natural human aspirations to knowledge and objectivity, 'philosophical realism' now names an intellectual current

that ultimately serves only to corrode our conviction in the possibility of attaining either" (p. xv). For a somewhat revised position, see Putnam (1994).

16. Kuhn (1991), p. 104.
17. See Hoyningen-Huene (1989a/1993), pp. 267–271; and (1989b); see also Nola (1980); Sankey (1997), chaps. 2–4; and Ghins (1998).
18. Kuhn (1992), p. 120.
19. Reaffirming his opposition to the traditional, correspondence theory of truth, Kuhn sketches his position about the nature of truth in the context of the idea of variant lexical structures. Scientific theories are the source of alternative sets of taxonomic categories which are imposed on the world by theories. Such categories do not reflect reality: it is impossible for them to do so—they are, at most, ways of ordering experience (taxonomy). The sets imposed on the world depend on theories and vary with them. In a Kantian way (but with mobile categories), the taxonomic categories of the scheme provide a structure for possible experience. Thus we can say, following Kuhn's own image, that as theories change, the world changes with them. See Kuhn (1991) and Sankey (1997).
20. See Hoyningen-Huene (1989a), pp. 267–271; and (1989b); see also Sankey (1993) and (1997).
21. That is the reason why Kuhn, for example, uses the term "puzzle" instead of "problem" (a choice analogous to Wittgenstein's "game"): according to this view no room is left for any critical form of rationality that is independent from (let alone opposed to) the dominant epistemic practices in a given historical period. A crucial issue uniting Kuhn and the logical positivists is their utter refusal to link radical criticism, or a choice among competing normative systems, with anything "rational": in their eyes, rationality is not to be found in establishing or breaking a rule, but merely (*à la* Wittgenstein) in following it, or abiding by it. See Kuhn (1962), pp. 36–37; and—by contrast—*UQ*, p. 122; and Popper (1970), p. 53.
22. More than ever, the striking similarity with Carnap is evident. According to Carnap, internal questions can be answered by referring to the logical rules of a given linguistic framework. In this case we have genuine theoretical questions, to which the notions of "correct" or "incorrect," "true" or "false" clearly and unproblematically apply. Researchers sharing a given linguistic framework can engage in theoretically genuine disputes about such internal questions. On the contrary, external questions, essentially involving a choice among different linguistic frameworks, are not genuinely rational in this sense. For, in the latter case, we are confronted with questions of a purely pragmatic or instrumental character about the adequacy or appropriateness of a given framework, designed in view of a given aim. This means, in the first place, that answers to external questions cannot be assessed by appealing to dichotomies like "correct" or "incorrect," "true" or "false," but nearly always involve problems of degrees. Secondly, such a distinction implies that answers to external questions are necessarily relative to the goals individual researchers aim for—more cautious researchers, fearing to contradict themselves, could, for example, prefer the weaker rules of intuitionist logic, while those interested in a wider applicability of physics may opt for the more binding rules of classical logic. See Carnap (1928), (1934), (1935), (1936–1937), (1956), and (1963); for a discussion, see Reisch (1991), pp. 270–274.
23. Feyerabend (1995), p. 356.
24. Following William Bartley's suggestion, we could call the latter "Wittgensteinian problematic": see Bartley (1990), chaps. 14–15.

25. "Instead of being a faulty sort of deduction, induction is fundamental, defining science—just as deduction is fundamental, defining logic" (Bartley 1990, p. 219).

26. See Wittgenstein (1953), part 1, §66: "don't think, but look!"; "we may not advance any kind of theory. There must not be anything hypothetical in our considerations. We must do away with all explanation, and description alone must take its place" (§109); see also §89; and his (1969), §189. It is no accident that Wittgenstein had already used the term "paradigm" to refer to what rules in an activity very similar to Kuhn's "normal science": see his (1953), part 1, §50 and §54.

27. See Feyerabend (1970a). "Consolations for the Specialist" is the title of Feyerabend's critical remarks on Kuhn's paper, as delivered at the 1965 Bedford Colloquium in London; see Kuhn (1970). The affinities between Kuhn's position and Wittgenstein's are substantial: see, in particular, Wittgenstein (1921), 4.112; (1953), part 1, §109; and (1969), §189. To Wittgenstein, philosophy has no cognitive function—rather, it has a "therapeutic" function; see his (1953), part 1, §109, §133, and §255. The descriptive task which characterizes philosophy concerns the rules governing the use of our language, that is, the grammar of the terms that constitute it: "description" refers to the description of language games, and it aims at showing the rules of those games and hence the structures which characterize them. Concerning rules, and not facts, description has an exemplary value. There is a close parallelism between the role of Kuhn's "exemplars" ("or exemplary problem solutions") and Wittgenstein's "examples": see Kuhn (1970) and (1974) (both of which were written in 1969, a crucial year for Kuhn's philosophical development); and Wittgenstein (1953), part 1, §71, §77, and §133.

28. As Feyerabend immediately spotted, upon reading a draft of *The Structure of Scientific Revolutions*, "You present your material in such a manner that (at least for the periods following the introduction of paradigms) history seems to satisfy the principle that alles wirkliche ist vernünftig so that evaluations can then be directly derived from historical study" (Feyerabend 1995, p. 355). In his (1970b) Feyerabend sought to rehabilitate Ernst Mach's philosophy, and particularly his "knowledge without foundations," to criticize much of contemporary philosophy of science, and especially the very many epistemologies that had been erected "in a spirit of conformism" and that had lost any chances of making an effective contribution to our knowledge of the world. As a matter of fact, in the past few years the academic world witnessed the increasing proliferation of disciplines and subdisciplines: more than ever, specialists of one discipline tend to lose sight of what is going on in the others, out of lack of interest or of actual acquaintance with the technical knowledge required to keep pace with the ongoing debates in one discipline or another. Of course, increasing specialization is not wrong in itself: proliferations of problems and their studies are all to the good. But there may be a danger here. For, on the one hand, disciplines tend to propagate, rather than progress: the issues discussed no longer address wide-ranging problems but focus on details, consistently avoiding generalization. On the other hand, practitioners of disciplines and subdisciplines found ever more specialized journals, meet at narrow-focused conferences and speak their technical languages—that is, they tend to isolate themselves from others and evade interdisciplinary confrontation. There are many reasons for this tendency, of course. But I do think Kuhn's philosophy, with its emphasis on the proliferation of allegedly incommensurable specialties and subspecialties as the only measure of scientific and philosophical progress, may be largely responsible for it. Kuhn-style proliferation is evasive, especially of controversy, and may

lead to intellectual stagnation. A way to avoid that, I suggest, is the resistance to the taboo of going beyond one's specialization—after all, those stagnate who have no intent to move forward. Specialism, in this respect, is the worst antidote.

29. As Kant warned, "Laziness and cowardice are the reasons why such a large portion of men, even when nature has long emancipated them from alien guidance (*naturaliter maiorennes*), nevertheless gladly remain immature for life. For the same reasons, it is all too easy for others to set themselves up as their guardians. It is so convenient to be immature! If I have a book to have understanding in place of me, a spiritual adviser to have a conscience for me, a doctor to judge a diet for me, and so on, I need not make any efforts at all. I need not think, so long as I can pay; others will soon enough take the tiresome job for me" (Kant 1784/1970, p. 54).

30. This issue has been particularly raised by Steve Fuller. In Fuller's view, the effect of *The Structure of Scientific Revolutions* was that philosophers of science relinquished their critical attitudes towards science and turned from treating science prescriptively to treating it merely descriptively, thus becoming merely "underlabourers" (Fuller 2000, pp. 260–265). By the same token, social scientists used Kuhn's magnum opus to reinvent themselves as "real scientists," thereby eliminating their critical and political stance towards the technological and economic dimension of science. "The surest historical measure of progress in Kuhn's account of science," writes Fuller, "is the increased specialisation of disciplinary research agendas. In my own less charitable gaze, this appears as a cancerous growth of mutually impenetrable jargons that obstructs the search for more holistic—if not 'unified'—understanding of reality, a project that would carry on the spirit of *episteme, scientia, Wissenschaft*, and science, as these concepts have been understood throughout most of the Western tradition" (ibid., p. 13). Indeed, according to Fuller's harsh reading, Kuhn's influence has been permeating the past quarter-century philosophy of science both in structure and in content—allowing, on the one hand, the fragmentation of the field and, on the other, the disconnection of the notion of scientific progress from any substantive ends science may be seen as pursuing.

31. Despite Kuhn's talk of revolutions and acknowledgment of the existence and necessity of revolutionary periods in the growth of scientific knowledge, the notions of paradigm (or lexicon) and normal science dominate Kuhn's picture: they are required if we want to progress, they constitute the very essence of scientific enterprise.

32. See Notturno (2000), chap. 10, especially pp. 238–239.

33. For a reconstruction of the various steps that led to the confrontation, as well as an outline of the papers delivered and the discussion that followed, see Gattei (forthcom.), chap. 2.

34. Kuhn (1970), p. 2.

35. Kuhn (1970), p. 2.

36. Ibid.

37. The original title is preserved in the draft of Kuhn's paper, a copy of which is in Popper Archive (80.9); see also Kuhn (1970), pp. 2–3.

38. Kuhn (1962, 1970), p. 175.

39. See Kuhn (1970), pp. 4–5.

40. Ibid., p. 6.

41. Kuhn distinguishes between the world in itself and the world of phenomena: in his eyes the reality which is usually addressed in everyday or scientific contexts is a world of phenomena, not the (single) world of phenomena, and certainly not the world in itself. In the web of similarity and dissimilarity

relations that constitute a given world of phenomena, we find a blend of objective and genetically subjective elements (not at the individual level, but at the social one: if we want to find an idealistic element in Kuhn's idea of reality, this has a social, not an individual nature). When examining a web of this kind we cannot separate those two moments. As a consequence, it is not possible to "purify" the world of phenomena from its subjective components, in order to achieve a "pure" picture of the objective elements, absolute reality or the world in itself. On the contrary, the concrete properties of the world in itself are inaccessible to us: even if we feel the resistance that world offers against our epistemic attempts, we are not in the position to grasp this very resistance in itself.

42. Kuhn (1970), p. 21. In a letter to Popper dated June 30, 1965, in Popper Archive (317.17), Kuhn claims that the line Popper seems sometimes to be drawing between history on the one hand and psychology on the other, appears to be arbitrary: Kuhn believes that we can (and ultimately must) understand the nature of the growth of scientific knowledge through the understanding of the nature of the community responsible for its creation, production, and protection.

43. Kuhn (1970), p. 6. Severity of tests and a problem-solving tradition: both characterize science, according to Popper. That is why Popper's and Kuhn's lines of demarcation so often coincide—but such a coincidence, Kuhn hastens to point out, is "only in their outcome; the process of applying them is very different, and it isolates distinct aspects of the activity about which the decision—science or non-science—is to be made" (ibid., p. 7). The example is astrology: whereas Popper excludes it from sciences for the way in which its practitioners explained their failures, Kuhn excludes it because though astrologers "had rules to apply, they had no puzzles to solve and therefore no science to practise" (ibid., p. 9). What is lacking, in other words, is a puzzle-solving (or research) tradition, that is, the kind of activity that "normally" characterizes all sciences acknowledged as such: "To rely on testing as the mark of a science it to miss what scientists mostly do and, with it, the most characteristic feature of their own enterprise" (ibid., p. 10).

44. John Watkins suggested this very parallel in 1961 after reading the manuscript of *The Structure of Scientific Revolutions*: see Watkins (1970), p. 26.

45. See Popper's "Revolution and Continuity in Science," in Popper Archive (120.11); as well as *RMC*, pp. 1144–1148. From the evolutionary point of view developed by Popper, the routine seems to be characterizing the way in which animals learn, or the way in which they adapt themselves to the environment. Man, on the contrary, by means of the invention of language—that has, among others, descriptive and argumentative functions—"has begun to replace routine more and more by critical approach" (*RMC*, p. 1146), and science is the most advanced application of the critical approach to the growth of knowledge. Popper sees in science, taken in an evolutionary context, "the conscious and critical form of an adaptive method of trial and error" (ibid., p. 1147): for this reason we can learn from our errors, in a permanently revolutionary process, constantly characterized by revolutions at various levels.

46. Popper is a firm supporter of dissent: "I am not an admirer of philosophical discipline" (*P1*, p. 7, where he also tells the story of the soldier who found that his whole battalion—except himself, of course—was out of step: "I constantly find myself in this entertaining position. And [. . .] I am content as long as enough members of the battalion are sufficiently out of step with one another"). He actually thinks it is possible to spot the secret of the flourishing of Greek philosophy, that every new generation produced

a new cosmology of surprising originality and profundity (see *CR*, chap. 5), exactly in the tradition of critical discussion. The possibility of fighting with words instead of swords is, for Popper, the very basis of our civilization, and particularly of its legal and parliamentary institutions, as well as the hallmark of scientific reason.

47. *BGE*, p. 136.
48. Popper (1970), p. 55; see also *CR*, pp. 247 and 312n1; and *RMC*, p. 984. "That it is desirable that a theory should be defended with a certain dogmatism, so that it is not knocked out too quickly before its resources have been explored, Popper has never denied; but such dogmatism is healthy only as long as there are other people who are not inhibited from criticizing and testing a tenaciously defended theory. If everyone were [. . .] to preserve the current theories of science against awkward results, then those theories would, according to Popper, lose their scientific status and degenerate into something like metaphysical doctrines" (Watkins 1970, p. 28). Toulmin agrees: "One cannot even label a false trail as such without exploring it some way first" (Toulmin 1961, p. 81).
49. Popper (1970), p. 55.
50. Ibid., p. 56.
51. Ibid.; see also *MF*, chap. 2; and *CR*, chap. 10. Popper's most vibrant and effective criticism of relativism is to be found in *OS2*, pp. 369–396.
52. Kuhn correctly highlights the existence, in science, of a community of professionals whose training has been mostly by indoctrination. We live within communities and are a product of a century-old tradition (if each time we had to start from the beginning, as the positivists wanted, it is reasonable to think that we would reach more or less the point Adam reached: our progress beyond him is due to the existence of a tradition); we grew within a cultural framework and we are in need of it. Popper is fully aware of this. What he (together with Watkins) dislikes, under the rubric of "normal science" is mental rigidity and, contrary to Kuhn, he wishes to fight it. This is the meaning of the expression (somewhat *à la* Leon Trotzky) "revolution in permanence": although we are prisoners caught in the framework the tradition provided us, we can always try to pull down the walls of the prison and escape. All we need is the will to do that: see the notes in Popper Archive (75.5).
53. Popper (1970), p. 56. Popper spots this as the central point of his disagreement with Kuhn; see also Popper's letter to Kuhn, July 7, 1965, in Popper Archive (317.17).
54. Popper (1970), p. 56; see also Popper Archive (75.5); *OK*, pp. 215–216; and *RMC*, pp. 1148–1153.
55. See *OK*, chap. 3; and *P2*, pp. 113–130.
56. Popper's meaning for this word is clearly described in his *OS2*, pp. 369–396.
57. These are the words Popper uses in an unpublished typescript, in Popper Archive (75.5), p. 9; in his (1970), p. 56, the remark is slightly softened. This is both a logical difference, as it concerns the role played by truth in scientific research and in the appraisal and choice of different theories; and a metaphysical difference, as it concerns their different approaches to science and philosophy, the different solution they provide to the problem of rationality.
58. See Miller (2006), chaps. 3 and 7.
59. See Kuhn's own recollections in his last interview: Kuhn (1997).
60. Brown (1977), p. 67.
61. See Pera (1981), p. 3.
62. See Notturno (2000), pp. 116–119.
63. Miller (2006), p. 57.
64. See Notturno (2000), especially chap. 5.

65. That is why, for example, errors are often regarded as something to be avoided and not as something we can learn from.

66. This is the sense of Borges' poem at the opening of this book: the progress of knowledge is fuelled by open, critical dialogue; speakers only share a faith in confrontation as the only way, however difficult, to approach the truth; they appeal to reason alone, honestly, without any verbalisms; their arguments may abound in errors, but this does not matter, since errors can be corrected; as far as possible, they are free from dogma and irrationalism, or they strive to be; their goal is not to convince opponents, but to advance knowledge: truth is the only regulative idea.

67. Plato [1925], 202a.

68. See Artigas (1999); and Agassi (2003), chap. 1.3.

69. Feyerabend (1961), pp. 55–56.

NOTES TO CHAPTER 5

1. Sextus Empiricus [1933], Book I, ch. XII, p. 21.

2. Xenophanes, DK B 18; the translation is Popper's own: see *TWP*, p. 48.

3. Xenophanes, DK B 34, Popper's translation: *TWP*, p. 46.

4. *LSD*, p. 53.

5. *OS1*, p. 124.

6. See *OS1*, pp. 120–121; as well as Bartley (1962, 1984), pp. 110–111.

7. Bartley (1962, 1984), pp. 112–113; see also Bartley (1964).

8. Ibid., p. 113, emphasis suppressed.

9. "These early fideistic remarks are relatively unimportant; they play no significant role in Popper's early thought and none at all in his later thought, but are superfluous remnants of justificationism, out of line with the main thrust and intent of his methodology, empty baggage carried over from the dominant tradition" (Bartley 1990, p. 237). In the same line, Agassi tried to free Popper's philosophy from inconsistencies deriving from his views on the role of corroborations: see Agassi (1961) and (1968).

10. Rationality is located in criticism. "A rationalist becomes one who holds *everything*—including standards, goals, criteria, authorities, decisions, and *especially* any framework or way of life—open to criticism" (Bartley 1990, p. 238). Nothing is withheld from examination and review; any framework (contrary to Kuhn's paradigms or lexicons) is held rationally only to the extent that it is subjected to and survives criticism. See also Bartley (1962, 1984), pp. 118–125.

11. *OS2*, p. 231. See also *LSD*, p. 37 and p. 38: "Thus I freely admit that in arriving at my proposals I have been guided, in the last analysis, by value judgments and predilections. But I hope that my proposals may be acceptable to those who value not only logical rigour but also freedom from dogmatism; who seek practical applicability, but are even more attracted by the adventure of science, and by discoveries which again and again confront us with new and unexpected questions, challenging us to try out new and hitherto undreamed-of answers"; and *BGE*, p. 395: "We thus share the conventionalist standpoint that the ultimate foundation of all knowledge ought to be sought in an act of free determination, that is, in the setting of a goal that cannot, in its turn, be further rationally grounded. And this is, though in a different form, Kant's idea of the primacy of practical reason."

12. *OS2*, p. 225. The very realization of this would have saved much time for those who become prisoners of unending discussions about naive, sophisticated, or methodological falsificationism, whose overdetailed writings "may

or may not give an image similar to Sterne's detailed and chaotic but rather charming picture of inept provincial life" (Agassi 1971, p. 323). Moreover, if we ignore the ethical character of his reasoning, we risk reducing a remarkable system of ideas to a soulless body of unsolved problems.

13. Excerpts from the verbatim transcription of Popper's impromptu reply during the Workshop Seminar in Kyoto, November 12, 1992: in Artigas (1999), p. 33. Popper goes on to say that he is "sorry" for the discussion that emerged from Bartley's criticism, since it was an abstract discussion, which resulted in damage to the moral attitude of critical rationalism (ibid.). Also, he felt as if Bartley's attempt bore the character of a definition, thus leading to endless philosophical arguments about its adequacy: see *MF*, pp. XII–XIII.

14. Logical reasoning occupies a central position in Popper's epistemology, but his critical attitude involves something more and different. His logical reasoning is a consequence of his ethical experiences, and its meaning is a part of a wider problem involving the ethical responsibility of human beings in general. For Bartley's view of the ethical dimension of choices and the suffering involved in the practice of reason, see his (1990), pp. 249–250 and 264–265.

15. *MF*, p. XII.

16. *CR*, p. 357.

17. *OS2*, p. 240. As Malachi Hacohen rightly observes, "Bartley's critique was a major advance for critical rationalism, but, historically, it *was* Popper's irrational commitment to rationalism that gave rise to his philosophy. Bartley wisely disposed of the justificationist ladder once he had seen the world aright" (Hacohen 2000, p. 519n259). I agree with Hacohen: it was Popper's irrational faith in reason that allowed him to build the edifice of critical rationalism—but I suggest it was a moral, rather than justificationist, ladder. As such, it cannot be disposed of, or else we would find ourselves in the position of giving up what we cannot do without. The situation is similar to the one described by Kant in the "Introduction" to the *Critique of Pure Reason*. There he exposes the illusion of Platonic idealism: "The light dove, cleaving the air in her free flight, and feeling its resistance, might imagine that its flight would be still easier in empty space. It was thus that Plato left the world of the senses, as setting too narrow limits to the understanding, and ventured out beyond it on the wings of the ideas, in the empty space of the pure understanding. He did not observe that with all his efforts he made no advance—meeting no resistance that might, as it were, serve as a support upon which he could take a stand, to which he could apply his powers, and so set his understanding in motion" (Kant 1781/2003, p. 47). Just as for Kant's "light dove" air is not an annoying friction—it is what renders its very flight possible—so, for Popper, an initial faith in reason, however irrational it may be, is the precondition for dialogue and criticism, that is, for the very use of reason.

18. Hacohen (2000), p. 541.

19. Ibid., p. 545. We must however keep in mind that open society does not equate with democracy.

20. Believing in reason is not sufficient; we must put it into action and practice it—particularly with people whose views and styles are different from, and therefore more of a challenge to, our own. That is the first way we have to respect others, that is, to allow them to make a difference to us, to affect our views and to have an impact on our own lives.

21. Popper's invitation calls us back to Kant's words: "*Enlightenment is man's emergence from his self-incurred immaturity. Immaturity is the inability to use one's own understanding without the guidance of another. The imma-*

turity is *self-incurred* if its cause is not lack of understanding, but lack of resolution and courage to use it without the guidance of another. The motto of enlightenment is therefore: *Sapere aude!* Have courage to use your own understanding!" (Kant 1784/1970, p. 54).

22. *OS2*, p. 232.
23. See *UQ*, pp. 193–198.
24. As Popper said in his later years, "we should not shrink from waging war for peace" (*ALPS*, p. 119).
25. See *BGE*, book II, chap. 7; as well as *LSD*, chap. 2.
26. See *LSD*, pp. 280–281.
27. See, for example, Agassi (1982); (1988), chap. 43; and (1990), chap. 1; as well as *MF*, pp. xii–xiii. Important choices in life, such as philosophical viewpoints and ethical standards, are usually not the result of argument or logical reflection any more than scientific theories are the result of sense observation.
28. Indeed, we can read Popper's theory of rationality and freedom as an attempt to maintain a coherence between the political and the epistemological.
29. Watkins (1974), pp. 406–407.
30. Milton (1667, 1674), book III, p. 108. Of course, not deciding is already a decision, though of the worst kind.
31. For sure, it may lead to these things—but it also may not, and faith in reason may become important precisely when it does not.
32. Feyerabend (1961), p. 56.

Bibliography

AGASSI, Joseph
1961   "The Role of Corroboration in Popper's Methodology." *Australasian Journal of Philosophy* 39 (1961): 81–91. Reprinted in Joseph Agassi, *Science in Flux*. Dordrecht-Boston: D. Reidel Publishing Company, 1975, 40–50.
1964a   "The Nature of Scientific Problems and Their Roots in Metaphysics." In *The Critical Approach to Science and Philosophy: In Honor of Karl R. Popper*, edited by Mario Bunge, 189–211. New York: The Free Press, 1964. Reprinted in Joseph Agassi, *Science in Flux*. Dordrecht-Boston: D. Reidel Publishing Company, 1975, 208–233.
1964b   "The Confusion Between Physics and Metaphysics in Standard Histories of Science." In *Ithaca: Proceedings of the Xth International Congress for the History of Science*, edited by Henry Guerlac, 231–238, 249–250. Paris: Hermann, 1964. Reprinted in Joseph Agassi, *Science in Flux*. Dordrecht-Boston: D. Reidel Publishing Company, 1975, 270–281.
1968   "Science in Flux. Footnotes to Popper." In *In Memory of Norwood Russell Hanson*, edited by Robert S. Cohen and Marx W. Wartofsky, 293–323. Dordrecht: D. Reidel Publishing Company, 1968. Reprinted in Joseph Agassi, *Science in Flux*. Dordrecht-Boston: D. Reidel Publishing Company, 1975, 9–39.
1971   "Tristram Shandy, Pierre Menard, and All That: Comments on *Criticism and the Growth of Knowledge*." *Inquiry* 14 (1971): 152–164. Reprinted as "Kuhn and His Critics: Rational Reconstruction of the Ant Heap," in Agassi (1988), 315–328.
1975   "Questions of Science and Metaphysics." In *Science in Flux*, 240–269. Dordrecht-Boston: D. Reidel Publishing Company, 1975.
1982   "In Search of Rationality." In *In Pursuit of Truth: Essays on the Philosophy of Karl Popper on the Occasion of His 80th Birthday*, edited by Paul Levinson, 237–248. Atlantic Heights: Humanities, 1982.
1988   *The Gentle Art of Philosophical Polemics: Selected Reviews and Comments*. La Salle, Illinois: Open Court Publishing Company, 1988.
1990   *The Siblinghood of Humanity: An Introduction to Philosophy*. Delmar: Caravan Books, 1990, 1991[2].
1993   *A Philosopher's Apprentice: In Karl Popper's Workshop*. Amsterdam-Atlanta: Rodopi, 1993.
2003   *Science and Culture*. Dordrecht-Boston-London: Kluwer Academic Publishers, 2003.

ARTIGAS, Mariano
1999   *The Ethical Nature of Karl Popper's Theory of Knowledge*. Berne: Peter Lang AG, 1999.

AXTELL, Guy S.
1993   "In the Tracks of the Historicist Movement. Re-assessing the Carnap-Kuhn Connection." *Studies in History and Philosophy of Science* 24 (1993): 119–146.

BARTLEY, William W., III
1962   *The Retreat to Commitment.* New York: Alfred A. Knopf, 1962. Revised edition, La Salle, Illinois: Open Court, 1984.
1964   "Rationality Versus the Theory of Rationality." In *The Critical Approach to Science and Philosophy*, edited by Mario Bunge, 3–31. New York: The Free Press, 1964.
1970   "Die österreichische Schulreform als die Wiege der modernen Philosophie." *Club Voltaire. Jahrbuch für kritische Aufklärung* IV (1970): 349–366.
1974   "Theory of Language and Philosophy of Science as Instruments of Educational Reform: Wittgenstein and Popper as Austrian Schoolteachers." In *Methodological and Historical Essays in the Natural and Social Sciences*, edited by Robert S. Cohen and Marx W. Wartofsky, 307–337. Dordrecht: D. Reidel Publishing Company, 1974.
1976   "On Imre Lakatos." In *Essays in Memory of Imre Lakatos*, edited by Robert S. Cohen, Paul K. Feyerabend, and Marx W. Wartofsky, 37–38. Dordrecht-Boston: D. Reidel Publishing Company, 1976.
1990   *Unfathomed Knowledge, Unmeasured Wealth: On Universities and the Wealth of Nations.* La Salle, Illinois: Open Court Publishing Company, 1990.

BERKSON, William
1976   "Lakatos One and Lakatos Two: An Appreciation." In *Essays in Memory of Imre Lakatos*, edited by Robert S. Cohen, Paul K. Feyerabend, and Marx W. Wartofsky, 39–54. Dordrecht-Boston: D. Reidel Publishing Company, 1976.

BERKSON, William, and John R. WETTERSTEN
1982   *Lernen aus dem Irrtum: Die Bedeutung von Karl Poppers Lerntheorie für die Psychologie und die Philosophie der Wissenschaft.* Hamburg: Hoffmann und Campe, 1982. English edition, *Learning from Error: Karl Popper's Psychology of Learning.* La Salle, Illinois: Open Court, 1984.

BROWN, Harold I.
1977   *Perception, Theory, and Commitment: The New Philosophy of Science.* Chicago: Precedent Publishing, 1977.

CARNAP, Rudolf
1928   *Der logische Aufbau der Welt.* Berlin: Weltkreis, 1928. Reprint, Berlin: Felix Meiner, 1962². Translated by Rolf A. George as *The Logical Structure of the World and Pseudoproblems in Philosophy*, 1–300 (London: Routledge & Kegan Paul, 1967).
1932a   "Überwindung der Metaphysik durch logische Analyse der Sprache." *Erkenntnis* 2 (1932): 219–241. Translated by Arthur Pap as "The Elimination of Metaphysics Through Logical Analysis of Language." In *Logical Positivism*, edited by Alfred J. Ayer, 60–81. Glencoe, Illinois: The Free Press, 1959.
1932b   "Die physikalische Sprache als Universalsprache der Wissenschaft." *Erkenntnis* 2 (1932): 432–465. Translated by Max Black as *The Unity of Science* (London: Kegan Paul, Trench, Trubner & Co., 1934).

1933a   "Psychologie in physikalischer Sprache." *Erkenntnis* 3 (1933): 107–142. Translated by George Schick as "Psychology in Physical Language." In *Logical Positivism*, edited by Alfred J. Ayer, 165–198. Glencoe, Illinois: The Free Press, 1959.

1933b   "Über Protokollsätze." *Erkenntnis* 3 (1933): 215–228.

1934   *Logische Syntax der Sprache.* Wien: Julius Springer, 1934. Translated by Amethe Smeaton (Countess von Zeppelin), revised by Olaf Helmer as *The Logical Syntax of Language* (London: Kegan Paul, Trench, Trubner & Co., 1937).

1935   *Philosophy and Logical Syntax.* London: Kegan Paul, 1935.

1936–37   "Testability and Meaning." *Philosophy of Science* 3 (1936): 419–471 and *Philosophy of Science* 4 (1937): 1–40. Reprinted as *Testability and Meaning.* New Haven: Yale University Press, 1950.

1956   "The Methodological Character of Theoretical Concepts." In *The Foundations of Science and the Concepts of Psychology and Psychoanalysis*, edited by Herbert Feigl and Michael Scriven, 38–76. Minneapolis: University of Minnesota Press, 1956.

1963   "Intellectual Autobiography." In *The Philosophy of Rudolf Carnap*, edited by Paul A. Schilpp, 3–84. La Salle, Illinois: Open Court Publishing Company, 1963.

DUHEM, Pierre
1906   *La théorie physique. Son objet et sa structure.* Paris: Chevalier & Rivière, 1906. Reprinted as *La théorie physique: son objet—sa structure.* Paris: Marcel Rivière, 1914². Translated by Philip P. Wiener as *The Aim and Structure of Physical Theory* (Princeton: Princeton University Press, 1954).

EARMAN, John
1993   "Carnap, Kuhn, and the Philosophy of Scientific Methodology." In *World Changes: Thomas Kuhn and the Nature of Science*, edited by Paul Horwich, 9–36. Cambridge, Massachusetts-London: The MIT Press, 1993.

EINSTEIN, Albert
1922   "Geometrie und Erfahrung." *Preußische Akademie der Wissenschaften, Physikalisch-mathematische Klasse, Sitzungsberichte* 1921, no. 1 (1922): 123–130. Translated by G. B. Jeffery and W. Perrett as "Geometry and Experience." In *Sidelights of Relativity*, 27–56. London: Methuen & Co., 1922.

FEYERABEND, Paul K.
1961   *Knowledge Without Foundations.* Oberlin, Ohio: Oberlin College, 1961. Reprinted in *Knowledge, Science and Relativism: Philosophical Papers*, vol. 3, edited by John M. Preston, 50–77. Cambridge: Cambridge University Press, 1999.

1970a   "Consolations for the Specialist." In *Criticism and the Growth of Knowledge*, edited by Imre Lakatos and Alan E. Musgrave, 197–230. Cambridge: Cambridge University Press, 1970.

1970b   "Philosophy of Science: A Subject with a Great Past." In *Historical and Philosophical Perspectives of Science*, edited by Roger H. Stuewer, 172–183. Minneapolis: University of Minnesota Press, 1970. Reprinted in *Knowledge, Science and Relativism: Philosophical Papers*, vol. 3, edited by John M. Preston, 127–137. Cambridge: Cambridge University Press, 1999.

1978   "Kuhns Struktur wissenschaftlicher Revolutionen. Ein Trostbüchlein
          für Spezialisten?" In *Der Wissenschaftstheoretische Realismus und die
          Autorität der Wissenschaften: Ausgewählte Schriften*, 153–204. Braun-
          schweig: Friedrich Vieweg & Sohn, 1978.
1995   "Two Letters of Paul Feyerabend to Thomas S. Kuhn on a Draft of *The
          Structure of Scientific Revolutions*." Edited by Paul Hoyningen-Huene.
          Studies in History and Philosophy of Science 26 (1995): 353–387.
2006   "More Letters by Paul Feyerabend to Thomas S. Kuhn on Proto-Struc-
          ture." Edited by Paul Hoyningen-Huene. *Studies in History and Phi-
          losophy of Science* 37 (2006): 610–632.

FREGE, F. L. Gottlob
1919   "Der Gedanke. Eine Logische Untersuchung." *Beiträge zur Philosophie
          des deutschen Idealismus* I (1918–1919): 58–77. Translated by R. H.
          Stoothoff as "Thoughts." In *Logical Investigations*, edited with a pref-
          ace by Peter T. Geach, 1–30. Oxford: Basil Blackwell, 1977.

FRIEDMAN, Michael
1991   "The Re-evaluation of Logical Positivism." *The Journal of Philosophy*
          88 (1991): 505–519. Reprinted as "Introduction" in *Reconsidering Log-
          ical Positivism*, 1–14. Cambridge: Cambridge University Press, 1999.
1992   "Philosophy and the Exact Sciences: Logical Positivism as a Case
          Study." In *Inference, Explanation, and Other Frustrations. Essays in
          the Philosophy of Science*, edited by John Earman, 84–98. Los Angeles:
          University of California Press, 1992.
1993   "Remarks on the History of Science and the History of Philosophy." In *World
          Changes: Thomas Kuhn and the Nature of Science*, edited by Paul Horwich,
          37–54. Cambridge, Massachusetts-London: The MIT Press, 1993.
2001   *Dynamics of Reason*. Stanford: CSLI Publications, 2001.
2003   "Kuhn and Logical Empiricism." In *Thomas Kuhn*, edited by Thomas
          Nickles, 19–44. Cambridge: Cambridge University Press, 2003.

FULLER, Steve W.
2000   *Thomas Kuhn: A Philosophical History of Our Times*. Chicago-Lon-
          don: The University of Chicago Press, 2000.

GATTEI, Stefano
2004   "Karl Popper's Philosophical Breakthrough." *Philosophy of Science* 71
          (2004): 448–466.
forthcom. *Thomas Kuhn's "Linguistic Turn" and the Legacy of Logical Positiv-
          ism: Incommensurability, Rationality and the Search for Truth*. Alder-
          shot: Ashgate, forthcoming.

GHINS, Michel
1998   "Kuhn: Realist or Antirealist?" *Principia* 2 (1998): 37–59.

GILLIES, Donald A.
1993   *The Philosophy of Science in the Twentieth Century: Four Central
          Themes*. Oxford: Blackwell Publishers, 1993.

HACOHEN, Malachi H.
2000   *Karl Popper—The Formative Years, 1902–1945: Politics and Philoso-
          phy in Interwar Vienna*. New York: Cambridge University Press, 2000.

HARDING, Sandra G. (ed.)
1976   *Can Theories Be Refuted? Essays on the Duhem-Quine Thesis.* Dordrecht-Boston: D. Reidel Publishing Company, 1976.

HARRIS, John H.
1974   "Popper's Definitions of Verisimilitude." *The British Journal for the Philosophy of Science* 25 (1974): 160–166.

HELMHOLTZ, Hermann von
1866   "Über die tatsächlichen Grundlagen der Geometrie." *Verhandlungen des naturhistorisch-medicinischen Vereins zu Heidelberg* 4 (1866): 197–202.
1868   "Über die Tatsachen, die der Geometrie zum Grunde liegen." *Nachrichten von der Kgl. Gesellschaft der Wissenschaften und der Georg-Augusts-Universität aus dem Jahre 1868* 9 (1868): 193–221. Translated by Malcom F. Lowe as "On the Facts underlying Geometry." In *Epistemological Writings*, edited by Robert S. Cohen and Yehuda Elkana, 39–58. Dordrecht-Boston: D. Reidel Publishing Company, 1977.
1876   "Über den Ursprung und die Bedeutung der geometrischen Axiome." In Hermann von Helmholtz, *Populare Wissenschaftliche Vortrage*, vol. 3, Braunschweig: Vieweg und Sohn, 1876, pp. 21–54. Translated by Malcom F. Lowe as "On the Origin and Significance of the Axioms of Geometry." In *Epistemological Writings*, edited by Robert S. Cohen and Yehuda Elkana, 1–26. Dordrecht-Boston: D. Reidel Publishing Company, 1977.

HOYNINGEN-HUENE, Paul
1989a *Die Wissenschaftsphilosophie Thomas S. Kuhns: Rekonstruktion und Grundlagenprobleme.* Braunschweig: Friedrich Vieweg & Sohn, 1989. Translated by Alexander T. Levine as *Reconstructing Scientific Revolutions: Thomas S. Kuhn's Philosophy of Science* (Chicago-London: The University of Chicago Press, 1993).
1989b "Idealist Elements in Thomas Kuhn's Philosophy of Science." *History of Philosophy Quarterly* 6 (1989): 393–401.

IRZIK, Gürol
2002   "Carnap and Kuhn: a Belated Encounter." In *In the Scope of Logic, Methodology and Philosophy of Science. Volume Two of the 11th International Congress of Logic, Methodology and Philosophy of Science, Cracow, August 1999*, edited by Peter Gärdenfors, Jan Woleński, and Katarzyna Kijana-Placek, 603–620. Dordrecht-Boston-London: Kluwer Academic Publishers, 2002.

IRZIK, Gürol, and Teo GRÜNBERG
1995   "Carnap and Kuhn: Arch Enemies or Close Allies?" *The British Journal for the Philosophy of Science* 46 (1995): 285–307.

JARVIE, Ian C., and Joseph AGASSI
1973   "Magic and Rationality Again." *The British Journal of Sociology* 24 (1973): 236–245. Reprinted in *Rationality: The Critical View*, edited by Joseph Agassi and Ian C. Jarvie, 385–394. Dordrecht-Boston-Lancaster: Martinus Nijhoff Publishers, 1987.

KANT, Immanuel
1781   *Kritik der reinen Vernunft*. Riga: Hartknoch, 1781, 1787². Translated
        by Norman Kemp Smith as *Critique of Pure Reason* (London: Mac-
        millan, 1929; reprint, New York: Palgrave Macmillan, 2003³).
1784   "Beantwortung der Frage: Was ist Aufklärung?" *Berlinische Monats
        schrift* IV (1784): 481–494. Translated by Hugh B. Nisbet as "An Answer
        to the Question: 'What is Enlightenment?'" In *Kant's Political Writings*,
        edited with an introduction by Hans S. Reiss, 54–60. Cambridge: Cam-
        bridge University Press, 1970, 1991².

KUHN, Thomas S.
1962   *The Structure of Scientific Revolutions*. Chicago-London: The Univer-
        sity of Chicago Press, 1962, 1970².
1970   "Logic of Discovery or Psychology of Research?" In *Criticism and the
        Growth of Knowledge*, edited by Imre Lakatos and Alan E. Musgrave,
        1–23. Cambridge: Cambridge University Press, 1970.
1974   "Second Thoughts on Paradigms" in *The Structure of Scientific Theo-
        ries*, edited by Frederick Suppe, 459–482. Urbana-Chicago-London:
        University of Illinois Press, 1974, 1977².
1989   "Possible Worlds in History of Science." In *Possible Worlds in Humani-
        ties, Arts and Sciences*, edited by Sture Allén, 9–32. Berlin: Walter de
        Gruyter, 1989. Reprinted in *The Road Since Structure: Philosophical
        Essays, 1970–1993, With an Autobiographical Interview*, edited by
        James Conant and John Haugeland, 58–86. Chicago-London: The Uni-
        versity of Chicago Press, 2000.
1991   "The Road Since Structure." In *PSA 1990. Proceedings of the 1990
        Biennial Meeting of the Philosophy of Science Association*, vol. 2,
        edited by Arthur Fine, Micky Forbes, Linda Wessels, 3–13. East Lan-
        sing, Michigan: Philosophy of Science Association, 1991. Reprinted in
        *The Road Since Structure: Philosophical Essays, 1970–1993, With an
        Autobiographical Interview*, edited by James Conant and John Hauge-
        land, 90–104. Chicago-London: The University of Chicago Press,
        2000.
1992   *The Trouble with the Historical Philosophy of Science, An Occa-
        sional Publication of the Department of the History of Science*.
        Cambridge, Massachusetts: Harvard University, 1992. Reprinted in
        *The Road Since Structure: Philosophical Essays, 1970–1993, With
        an Autobiographical Interview*, edited by James Conant and John
        Haugeland, 105–120. Chicago-London: The University of Chicago
        Press, 2000.
1993   "Afterwords." In *World Changes: Thomas Kuhn and the Nature of
        Science*, edited by Paul Horwich, 311–341. Cambridge, Massachusetts-
        London: The MIT Press, 1993. Reprinted in *The Road Since Structure:
        Philosophical Essays, 1970–1993, With an Autobiographical Inter-
        view*, edited by James Conant and John Haugeland, 224–252. Chicago-
        London: The University of Chicago Press, 2000.
1997   "'A Physicist who Became a Historian for Philosophical Purposes': A
        Discussion Between Thomas S. Kuhn and Aristides Baltas, Kostas Gav-
        roglu, Vasso Kindi." *Neusis* 6 (1997): 145–200. Reprinted as "Discus-
        sion with Thomas S. Kuhn." In *The Road Since Structure: Philosophical
        Essays, 1970–1993, With an Autobiographical Interview*, edited by
        James Conant and John Haugeland, 255–323. Chicago-London: The
        University of Chicago Press, 2000.

LAKATOS, Imre
1970 "Falsification and the Methodology of Scientific Research Programmes." In *Criticism and the Growth of Knowledge*, edited by Imre Lakatos and Alan E. Musgrave, 91–195. Cambridge: Cambridge University Press, 1970.

LARVOR, Brendan
2003 "Why did Kuhn's *Structure of Scientific Revolutions* Cause a Fuss?" *Studies in History and Philosophy of Science* 34 (2003): 369–390.

MILLER, David W.
1974a "Popper's Qualitative Theory of Verisimilitude." *The British Journal for the Philosophy of Science* 25 (1974): 166–177.
1974b "On the Comparison of False Theories by their Bases." *The British Journal for the Philosophy of Science* 25 (1974): 178–188.
1976 "Verisimilitude Redeflated." *The British Journal for the Philosophy of Science* 27 (1976): 363–381.
1994 *Critical Rationalism: A Restatement and Defence*. Chicago-La Salle, Illinois: Open Court Publishing Company, 1994.
2006 *Out of Error: Further Essays on Critical Rationalism*. Aldershot: Ashgate Publishing Company, 2006.

MILTON, John
1667 *Paradise Lost: A Poem Written in Ten Books*. London: Peter Parker, Robert Boulter and Matthias Walker, 1667. Revised edition, *Paradise Lost: A Poem, in Twelve Books*. London: S. Simmons, 1674.

NEURATH, Otto
1932 "Soziologie im Physikalismus." *Erkenntnis* 2 (1932): 393–431.
1933 "Protokollsätze." *Erkenntnis* 3 (1932): 204–214. Translated by Robert S. Cohen and Marie Neurath as "Protocol Statements." In *Philosophical Papers 1913–1946*, edited and translated by Robert S. Cohen and Marie Neurath, with the assistance of Carolyn R. Fawcett, 91–99. Dordrecht-Boston-Lancaster: D. Reidel Publishing Company, 1983.
1935 "Pseudorationalismus der Falsifikation." *Erkenntnis* 5 (1935): 353–365. Translated by Robert S. Cohen and Marie Neurath as "Pseudorationalism of Falsification." In *Philosophical Papers 1913–1946*, edited and translated by Robert S. Cohen and Marie Neurath, with the assistance of Carolyn R. Fawcett, 121–131. Dordrecht-Boston-Lancaster: D. Reidel Publishing Company, 1983.

NEWTON-SMITH, William H.
1981 *The Rationality of Science*. London: Routledge & Kegan Paul, 1981.

NIINILUOTO, Ilkka
1987 *Truthlikeness*. Dordrecht-Boston: D. Reidel Publishing Company, 1987.

NOLA, Robert
1980 "'Paradigm Lost, or the World Regained'—An Excursion into Realism and Idealism in Science." *Synthese* 45 (1980): 317–350.

NOTTURNO, Mark A.
2000 *Science and the Open Society: The Future of Karl Popper's Philosophy*. Budapest: Central European University Press, 2000.

NOVALIS (Friedrich von Hardenberg)
1960   *Schriften, vol. 2: Das philosophische Werk I.* Edited by Richard Samuel, Hans-Joachim Mähl, and Gerhard Schulz. Stuttgart: W. Kohlhammer Verlag, 1960.

PERA, Marcello
1981   *Popper e la scienza su palafitte.* Rome-Bari: Laterza, 1981.

PLATO
1925   "Symposium." in *Lysis, Symposium, Gorgias.* Translated by Walter R. M. Lamb, 80–245. London-Cambridge, Massachusetts: Harvard University Press, 1925, 1983[2].

POINCARÉ, Jules-Henri
1898   "On the Foundations of Geometry." *The Monist* 9 (1898): 1–43.
1902   *La science et l'hypothèse.* Paris: Flammarion, 1902. Translated by George B. Halsted as *Science and Hypothesis* (New York: Dover Publications, 1952).
1908   *Science et méthode.* Paris: Flammarion, 1908. Translated by Francis Maitland as *The Value of Science* (New York: Dover Publications, 1952).

POPPER, Karl R.
1935   *Logik der Forschung: Zur Erkenntnistheorie der modernen Naturwissenschaft.* Vienna: Julius Springer, 1935. Translated by Karl Popper, Julius Freed, and Lan Freed as *The Logic of Scientific Discovery* (London: Hutchinson & Co., 1959).
1945   *The Open Society and Its Enemies, vol. I: The Spell of Plato, vol. II: The High Tide of Prophecy: Hegel, Marx, and the Aftermath.* London: Routledge & Kegan Paul, 1945, 1966[5].
1950   "Indeterminism in Quantum Physics and in Classical Physics." *The British Journal for the Philosophy of Science* 1 (1950): 117–133 and 173–195.
1957a   *The Poverty of Historicism.* London: Routledge & Kegan Paul, 1957.
1957b   "The Propensity Interpretation of the Calculus of Probability, and the Quantum Theory." In *Observation and Interpretation: A Symposium of Philosophers and Physicists*, edited by Stephan Körner, 65–70. New York: Academic Press, and London: Butterworths Scientific Publications, 1957.
1959   "The Propensity Interpretation of Probability." *The British Journal for the Philosophy of Science* 10 (1959–1960): 25–42.
1963   *Conjectures and Refutations: The Growth of Scientific Knowledge.* London: Routledge & Kegan Paul, 1963, 1989[5].
1970   "Normal Science and Its Dangers." In *Criticism and the Growth of Knowledge*, edited by Imre Lakatos and Alan E. Musgrave, 51–58. Cambridge: Cambridge University Press, 1970.
1972   *Objective Knowledge: An Evolutionary Approach.* Oxford: Clarendon Press, 1972, 1979[2].
1974a   "Autobiography of Karl Popper." In *The Philosophy of Karl Popper*, vol. 1, edited by Paul A. Schilpp, 1–181. La Salle, Illinois: Open Court Publishing Company, 1974. Revised edition, *Unended Quest: An Intellectual Autobiography.* London: Fontana/Collins, 1976, 1992[4].
1974b   "Replies to My Critics." In *The Philosophy of Karl Popper*, vol. 2, edited by Paul A. Schilpp, 961–1197. La Salle, Illinois: Open Court Publishing Company, 1974.
1975   "The Rationality of Scientific Revolutions." In *Problems of Scientific Revolution: Progress and Obstacles to Progress in the Sciences*, edited by Rom

Harré, 72–101. Oxford: Oxford University Press, 1975. Reprinted in *Scientific Revolutions*, edited by Ian Hacking, 80–106. Oxford: Oxford University Press, 1981.

1979   *Die beiden Grundprobleme der Erkenntnistheorie.* Edited by Troels E. Hansen. Tübingen: J. C. B. Mohr (Paul Siebeck), 1979, 1994[2].

1982a   *Postscript to The Logic of Scientific Discovery.* Vol. 2, *The Open Universe: An Argument for Indeterminism.* Edited by William W. Bartley III. London: Hutchinson, 1982.

1982b   *Postscript to The Logic of Scientific Discovery.* Vol. 3, *Quantum Theory and the Schism in Physics.* Edited by William W. Bartley III. London: Hutchinson, 1982.

1983   *Postscript to The Logic of Scientific Discovery.* Vol. 1, *Realism and the Aim of Science.* Edited by William W. Bartley III. London: Hutchinson, 1983.

1984   *Auf der Suche nach einer besseren Welt: Vorträge und Aufsätze aus dreißig Jahren.* Munich: R. Piper GmbH & Co., 1984. Translated by Laura J. Bennett as *In Search of a Better World: Lectures and Essays from Thirty Years* (London-New York: Routledge, 1994).

1990   *A World of Propensities.* Bristol: Thoemmes Antiquarian Books, 1990.

1994a   *The Myth of the Framework: In Defence of Science and Rationality.* Edited by Mark A. Notturno. London-New York: Routledge, 1994.

1994b   *Knowledge and the Body-Mind Problem: In Defence of Interaction.* Edited by Mark A. Notturno. London-New York: Routledge, 1994.

1994c   *Alles Leben ist Problemlösen: Über Erkenntnis, Geschichte und Politik.* Munich: R. Piper 1994. Translated by Patrick Camiller as *All Life is Problem-Solving* (London-New York: Routledge, 1999).

1998   *The World of Parmenides: Essays on the Presocratic Enlightenment.* Edited by Arne F. Petersen, with the assistance of Jørgen Mejer. London-New York: Routledge, 1998.

2006   *Frühe Schriften.* Edited by Troels Eggers Hansen. Tübingen: Mohr Siebeck, 2006.

POPPER, Karl R., and John C. ECCLES
1977   *The Self and Its Brain.* Berlin-New York: Springer, 1977, 1981[2].

PUTNAM, Hilary
1978   *Meaning and the Moral Sciences.* London-Boston: Routledge & Kegan Paul, 1978.

1981   *Reason, Truth and History.* Cambridge: Cambridge University Press, 1981.

1983   *Realism and Reason: Philosophical Papers, Volume 3.* Cambridge: Cambridge University Press, 1983.

1987   *The Many Faces of Realism: The Paul Carus Letters.* La Salle, Illinois: Open Court Publishing Company, 1987.

1990   *Realism with a Human Face.* Edited by James Conant. Cambridge, Massachusetts-London: Harvard University Press, 1990.

1994   "Sense, Nonsense, and the Senses: An Inquiry into the Powers of Human Mind." *The Journal of Philosophy* 91 (1994): 445–517.

QUINE, Willard V. O.
1951   "Two Dogmas of Empiricism." *Philosophical Review* 60 (1951): 20–43. Revised reprint in *From a Logical Point of View*, 20–46. Cambridge, Massachusetts: Harvard University Press, 1953, 1961[2].

REICHENBACH, Hans
1931   "Kausalität und Theoriebildung." *Erkenntnis* 1 (1931): 158–188. Partially translated by Maria Reichenbach as "Causality and Probability" in *Selected Writings 1909–1953*, vol. 2, edited by Maria Reichenbach and Robert S. Cohen, 332–344. Dordrecht-Boston-London: D. Reidel Publishing Company, 1978.
1933   "Bemerkung." *Erkenntnis* 3 (1933): 426–427.

REISCH, George A.
1991   "Did Kuhn Kill Logical Empiricism?" *Philosophy of Science* 58 (1991): 264–277.

SANKEY, Howard
1993   "Kuhn's Changing Concept of Incommensurability." *The British Journal for the Philosophy of Science* 44 (1993): 759–774. Reprinted in Sankey (1997), 21–34.
1997   *Rationality, Relativism and Incommensurability.* Aldershot: Ashgate, 1997.

SCHLICK, Moritz
1934   "Über das Fondament der Erkenntnis." *Erkenntnis* 4 (1934): 79–99. Translated by Peter Heath as "On the Foundation of Knowledge." In *Philosophical Papers. Volume II (1925–1936),* edited by Henk L. Mulder and Barbara F. B. van de Velde-Schlick, 370–387. Dordrecht-Boston-London: D. Reidel Publishing Company, 1979.
1936   "Meaning and Verification." *Philosophical Review* XLV (1936): 339–369. Reprinted in Moritz Schlick, *Philosophical Papers: Volume II (1925–1936),* edited by Henk L. Mulder and Barbara F. B. van de Velde-Schlick, Dordrecht-Boston-London: D. Reidel Publishing Company, 1979, pp. 456–481

SEXTUS EMPIRICUS
1933   *Outlines of Pyrrhonism.* Translated by Robert G. Bury. London-Cambridge, Massachusetts: Harvard University Press, 1933.

TER HARK, Michel
2002   "Between Autobiography and Reality: Popper's Inductive Years." *Studies in History and Philosophy of Science* 33 (2002): 79–103.
2004   *Popper, Otto Selz and the Rise of Evolutionary Epistemology.* Cambridge: Cambridge University Press, 2004.

TICHÝ, Pavel
1974   "On Popper's Definitions of Verisimilitude." *The British Journal for the Philosophy of Science* 25 (1974): 155–160.
1976   "Verisimilitude Redefined." *The British Journal for the Philosophy of Science* 27 (1976): 25–42.
1978   "Verisimilitude Revisited." *Synthese* 38 (1978): 175–196.

TOULMIN, Stephen E.
1961   *Foresight and Understanding: An Enquiry into the Aims of Science.* Bloomington: Indiana University Press, 1961.

VAIHINGER, Hans
1911   *Die Philosophie des als-ob. System der Theoretischen, praktischen und religiösen Fiktionen der Menschheit auf Grund eines idealistischen Positivismus.* Lipsia: F. Meiner, 1911, 1922[8].

WAISMANN, Friedrich
1930   "Logische Analyse der Wahrscheinlichkeitsbegriff." *Erkenntnis* 1 (1930–1931): 228–248. Translated by Hans Kaal as "A Logical Analysis of the Concept of Probability." In *Philosophical Papers*, edited by Brian McGuinness, 4–21. Dordrecht-Boston: D. Reidel Publishing Company, 1977.
1967   *Wittgenstein und der Wiener Kreis: Gespräche, aufgezeichnet von Friedrich Waismann.* Edited by Brian F. McGuinness. Oxford: Basil Blackwell, 1967. Translated by Joachim Schulte and Brian F. McGuinness as *Wittgenstein and the Vienna Circle: Conversations Recorded by Friedrich Waismann* (Oxford: Basil Blackwell, 1979).
2003   *The Voices of Wittgenstein: The Vienna Circle.* Edited by Gordon Baker. London-New York: Routledge, 2003.

WATKINS, John W. N.
1958   "Confirmable and Influential Metaphysics." *Mind* 67 (1958): 344–365.
1970   "Against 'Normal Science.'" In *Criticism and the Growth of Knowledge*, edited by Imre Lakatos and Alan E. Musgrave, 25–37. Cambridge: Cambridge University Press, 1970.
1974   "The Unity of Popper's Thought." In *The Philosophy of Karl Popper*, vol. 1, edited by Paul A. Schilpp, 371–412. La Salle, Illinois: Open Court Publishing Company, 1974.
1975   "Metaphysics and the Advancement of Science." *The British Journal for the Philosophy of Science* 26 (1975): 91–121.
1978   "Minimal Presuppositions and Maximal Metaphysics." *Mind* 87 (1978): 195–209.
1984   *Science and Scepticism.* Princeton: Princeton University Press, 1984.

WETTERSTEN, John R.
1985   "The Road through Würzburg, Vienna and Gottingen." *Philosophy of the Social Sciences* 15 (1985): 487–506.
1992   *The Roots of Critical Rationalism.* Amsterdam-Atlanta: Rodopi, 1992.
2005a   "New Insights on Young Popper." *Journal of the History of Ideas* 66 (2005): 603–631.
2005b   *Whewell's Critics. Have They Prevented Him from Doing Good?* Edited by James A. Bell. Amsterdam-New York: Rodopi, 2005.

WETTERSTEN, John R., and Joseph AGASSI
1991   "Whewell's Problematical Heritage," in *William Whewell: A Composite Portrait*, edited by Menachem Fisch, Simon Schaffer, 345–369. Oxford: Oxford University Press, 1991.

WITTGENSTEIN, Ludwig
1921   "Logisch-philosophische Abhandlung." *Annalen der Naturphilosophie* 14 (1921): 185–262. Translated by David F. Pears and Brian F. McGuinness, with an introduction by Bertrand Russell, as *Tractatus Logico-Philosophicus* (London: Routledge & Kegan Paul, 1961).
1953   *Philosophische Untersuchungen. Philosophical Investigations.* Translated by G. Elizabeth M. Anscombe. Edited by G. Elizabeth M. Anscombe and Rush Rhees. Oxford: Basil Blackwell, 1953.
1969   *On Certainty. Über Gewißheit.* Edited by G. Elizabeth M. Anscombe and Georg H. von Wright. Oxford: Basil Blackwell, 1969.

Name Index

Given the high number of occurrences, the name of Karl R. Popper has not been indexed.

Subject Index

Printed and bound by CPI Group (UK) Ltd, Croydon, CR0 4YY

01/11/2024

01782630-0013